RADIO WAVE PROPAGATION AND PARABOLIC EQUATION MODELING

D0879244

RADIO WAVE PROPAGATION AND PARABOLIC EQUATION MODELING

GÖKHAN APAYDIN

LEVENT SEVGI

IEEE PRESS

WILEY

Published by John Wiley & Sons, Inc., Hoboken, New Jersey.
Published simultaneously in Canada.

For general information on our other products and services or for technical support, please contact
our Customer Care Department within the United States at (800) 762-2974, outside the United States at
(317) 572-3993 or fax (317) 572-4002.

Wiley also publishes its books in a variety of electronic formats. Some content that appears in print may
not be available in electronic formats. For more information about Wiley products, visit our web site at
www.wiley.com.

Library of Congress Cataloging-in-Publication Data is available.

ISBN: 978-1-119-43211-1

Printed in the United States of America.

10 9 8 7 6 5 4 3 2 1

Table of Contents

PREFACE

Wave propagation in and through complex environments has long been a hot topic. One needs to understand/model/simulate electromagnetic wave propagation in order to establish a reliable communication link, to early detect targets in radar systems, to continuously cover an operational ground and air areas, etc. The same is also true for underwater acoustic waves. Establishing a communication link between a ship and a submarine, between two submarines, etc., and early detecting nuclear submarines in deep as well as shallow waters, etc., is a challenge. These are also critical in optical wave propagation through fiber cables and the atmosphere.

Realistic electromagnetic/acoustic/optical wave propagation models in and through complex environments have long been investigated and developed. One of the earliest and the most effective models, used in electromagnetic/acoustic/optical wave propagation, is the parabolic equation method. This book is a continuation of book *Parabolic Equation Methods for Electromagnetic Wave Propagation* written by Levy in 2000 and presents the application of analytical and numerical methods for wave propagation. The powerful numerical methods are given with several scenarios around us while considering the effects of environment on radio wave propagation.

This book is written for electrical, electronics, communication, computer engineers in industry, as well as for university students, researchers, and professors. The goal is to discuss fundamentals of electromagnetic wave propagation, especially on radio wave propagation, groundwave propagation, maritime communication, integrated maritime surveillance systems, submarine communication, defense industry, radar applications, etc. The topics listed in the contents are re-visited in terms of radio wave engineering. The book also introduces some simple MATLAB scripts for several well-known electromagnetic propagation problems.

The first chapter introduces some fundamental concepts of electromagnetic problems. Maxwell equations, transverse electric and magnetic models in guided wave representation, Fourier transform, validation, verification, and calibration are reviewed briefly. The second chapter presents the simplest propagation scenario used in analytical modeling to investigate wave propagation over flat Earth. The part is important for validation, verification, and calibration. Chapter 3 presents parabolic equation modeling based on the Fourier split-step and the finite-element methods. Simple MATLAB-based propagation tools are introduced. Systematic comparisons on some canonical test scenarios are performed, and the propagation tools are calibrated against the mathematically exact solutions. Atmospheric refractivity effects are also discussed at the end of this part. Accurate source modeling and validation, verification, and calibration of different propagators are discussed in Chapter 4. The SSPE, FEMPE, and MoM propagators are validated and calibrated at short ranges against the two-ray model over flat Earth and the single knife edge problem with the four-ray model. Irregular terrain modeling and impedance boundary modeling are presented in Chapter 5. Wave propagation inside a simple canonical structure, a parallel plate waveguide, is investigated in Chapter 6. Analytical solutions in terms of mode summation as well as image method are discussed and simple MATLAB scripts are developed. In Chapter 7, the classical rectangular waveguide problem is chosen for three-dimensional parabolic equation propagation. Propagation inside this waveguide can be exactly modeled in terms of modal summation and any given source distribution can be analytically represented as accurately as desired. The algorithms are developed, tested against analytically exact data, and calibrated here. Chapter 8 improves parabolic equation modeling with backward propagation, considered as two-way parabolic equation modeling. Simple MATLAB-based propagation tools which use two-way parabolic equation are developed and tested. Chapter 9 presents a novel software tool (PETOOL), developed in MATLAB with graphical user interface (GUI), for the analysis and visualization of radio wave propagation through the homogeneous and inhomogeneous atmosphere, by incorporating variable terrain effects with the aid of the two-way split-step algorithm employing wide-angle propagator. The last chapter presents a novel MATLAB software tool (FEMIX) for the analysis and visualization of surface-wave propagation over the irregular Earth's surface through a homogeneous and an inhomogeneous atmosphere.

Some of the contents of this book have been published in IEEE publications such as *IEEE Transactions on Antennas and Propagation*, and *IEEE Antennas and Propagation Magazine* for the last couple of years.

This book uses a MATLAB license under the MathWorks Book Program in developing book materials.

GÖKHAN APAYDIN, LEVENT SEVGI

ACRONYMS

ADIPE	alternate direction implicit parabolic equation
BC	boundary condition
CBC	Cauchy boundary condition
DBC	Dirichlet boundary condition
DCT	discrete cosine transform
DFT	discrete Fourier transform
DMFT	discrete mixed Fourier transform
DST	discrete sine transform
EM	electromagnetic
FDTD	finite-difference time domain
FEM	finite-element method
FEMIX	FEM-based PE algorithm with mixed paths
FEMPE	FEM-based parabolic equation
GO	geometric optics
GTD	geometric theory of diffraction

Radio Wave Propagation and Parabolic Equation Modeling, First Edition. By Gökhan Apaydin, Levent Sevgi
© 2017 by the Institute of Electrical and Electronic Engineers, Inc. Published 2017 by John Wiley & Sons, Inc.

GUI	graphical user interface
HFA	high-frequency asymptotics
FFT	fast Fourier transform
IM	image method
ITU	International Telecommunication Union
LOS	line of sight
MODSIM	modeling and simulation
MoM	method of moments
NBC	Neumann boundary condition
PDE	partial differential equation
PE	parabolic equation
PEC	perfect electric conductor
PF	propagation factor
PL	path loss
PO	physical optics
PWE	parabolic wave equation
PTD	physical theory of diffraction
RCS	radar cross-section
SSPE	split-step parabolic equation
TE	transverse electric
TEM	transverse electromagnetic
TM	transverse magnetic
TLM	transmission line matrix
UTD	uniform theory of diffraction
VV&C	validation, verification, and calibration
1D	one-dimensional
2D	two-dimensional
3D	three-dimensional

MATLAB CODES

CHAPTER 1

INTRODUCTION

1.1 Electromagnetic Problems and Classification

Electromagnetic (EM) problems are classified in terms of the equations describing them. The equations could be differential or integral or both. Most EM problems can be stated in terms of an operator equation

$$L\varphi = g \tag{1.1}$$

where L is an operator (differential, integral, or integro-differential), g is the known excitation or source, and φ is the unknown function to be determined. A typical example is an electrostatic problem involving Poisson's equation

$$-\nabla^2 V = \frac{\rho}{\varepsilon} \tag{1.2}$$

where $L = -\nabla^2$ is Laplacian operator, $g = \rho/\varepsilon$ is source term, and $\varphi = V$. In integral form, Poisson's equation is of the form

$$V = \int \frac{\rho dv}{4\pi\varepsilon r^2} \tag{1.3}$$

Radio Wave Propagation and Parabolic Equation Modeling, First Edition. By Gökhan Apaydin, Levent Sevgi
© 2017 by the Institute of Electrical and Electronic Engineers, Inc. Published 2017 by John Wiley & Sons, Inc.

where $L = \int \frac{dv}{4\pi r^2}$ is Laplacian operator, $g = V$ is source term, and $\varphi = \rho/\varepsilon$.

Electromagnetic problems involve linear, second-order differential equations. In general, a second-order partial differential equation (PDE) is given by

$$a\frac{\partial^2 \varphi}{\partial x^2} + b\frac{\partial^2 \varphi}{\partial x \partial y} + c\frac{\partial^2 \varphi}{\partial y^2} + d\frac{\partial \varphi}{\partial x} + e\frac{\partial \varphi}{\partial y} + f\varphi = g \qquad (1.4)$$

where the differential operator is

$$L = a\frac{\partial^2}{\partial x^2} + b\frac{\partial^2}{\partial x \partial y} + c\frac{\partial^2}{\partial y^2} + d\frac{\partial}{\partial x} + e\frac{\partial}{\partial y} + f. \qquad (1.5)$$

The coefficients, a, b, and c, in general are functions of x and y; they may also depend on φ itself, in which case the PDE is said to be nonlinear. A PDE in which $g(x, y)$ equals zero is termed homogeneous; it is inhomogeneous if $g(x, y)$ is not equal to zero.

A PDE, in general, can have both boundary values and initial values. PDEs whose boundary conditions (BCs) are specified are called steady-state equations. If only initial values are specified, they are called transient equations.

Any linear second-order PDE can be classified as elliptic, hyperbolic, or parabolic depending on the coefficients a, b, and c. The terms hyperbolic, parabolic, and elliptic are derived from the fact that the quadratic equation

$$ax^2 + bxy + cy^2 + dx + ey + f = 0 \qquad (1.6)$$

represents a hyperbola, parabola, or ellipse if $b^2 - 4ac$ is positive, zero, or negative, respectively.

In each of these categories, there are PDEs that model certain physical phenomena. Such phenomena are not limited to electromagnetics but extend to almost all areas of science and engineering. Thus the mathematical model specified here arises in problems involving heat transfer, boundary-layer flow, vibrations, elasticity, electrostatic, wave propagation, and so on.

Elliptic PDEs are associated with steady-state phenomena, that is boundary-value problems. Typical examples of this type of PDE include Laplace's equation

$$\frac{\partial^2 \varphi}{\partial x^2} + \frac{\partial^2 \varphi}{\partial y^2} = 0 \qquad (1.7)$$

and Poisson's equation

$$\frac{\partial^2 \varphi}{\partial x^2} + \frac{\partial^2 \varphi}{\partial y^2} = g(x, y) \qquad (1.8)$$

where in both cases $a = c = 1$, $b = 0$. An elliptic PDE usually models an interior problem, and hence the solution region is usually closed or bounded.

Hyperbolic PDEs arise in propagation problems. The solution region is usually open so that a solution advances outward indefinitely from initial conditions while always satisfying specified BCs. A typical example of hyperbolic PDE is the wave equation in one dimension

$$\frac{\partial^2 \varphi}{\partial x^2} = \frac{1}{v^2}\frac{\partial^2 \varphi}{\partial t^2} \qquad (1.9)$$

where $a = v^2$, $b = 0$, $c = 1$. If the time dependence is suppressed, the equation is merely the steady-state solution.

Parabolic PDEs are generally associated with problems in which the quantity of interest varies slowly in comparison with the random motions which produce the variations. The most common parabolic PDE is the diffusion (or heat) equation in one dimension

$$\frac{\partial^2 \varphi}{\partial x^2} = k \frac{\partial \varphi}{\partial t} \tag{1.10}$$

where $a = 1$, $b = c = 0$.

In hyperbolic and parabolic PDEs, the solution region is usually open. The initial conditions and BCs typically associated with parabolic equations resemble those for hyperbolic problems except that only one initial condition at $t = 0$ is necessary since the parabolic equation (PE) is only first-order in time. Also, parabolic and hyperbolic equations are solved using similar techniques, whereas elliptic equations are usually more difficult and require different techniques.

The type of problem represented by $L\varphi = g$ is said to be deterministic, since the quantity of interest can be determined directly. Another type of problem where the quantity is found indirectly is called non-deterministic or eigenvalue. The standard eigenproblem is of the form $L\varphi = \lambda\varphi$. A more general version is the generalized eigenproblem having the form $L\varphi = M\lambda\varphi$, where M, like L, is a linear operator for EM problems. Here, only some particular values of λ called eigenvalues are permissible; associated with these values are the corresponding solutions called eigenfunctions. Eigenproblems are usually encountered in vibration and waveguide problems where the eigenvalues λ correspond to physical quantities such as resonance and cutoff frequencies, respectively.

Our problem consists of finding the unknown function φ of a PDE. In addition to the fact that φ satisfies $L\varphi = g$ within a prescribed solution region R, φ must satisfy certain conditions on S, the boundary of R. Usually these BCs are

$$\varphi(\mathbf{r}) = 0, \ \mathbf{r} \text{ on } S, \ \text{(Dirichlet type)} \tag{1.11}$$

$$\frac{\partial \varphi(\mathbf{r})}{\partial n} = 0, \ \mathbf{r} \text{ on } S, \ \text{(Neumann type)}. \tag{1.12}$$

Here the normal derivative φ vanishes on S for Neumann type. Where a boundary has both, a mixed (Cauchy) boundary condition (CBC) is said to exist

$$\frac{\partial \varphi(\mathbf{r})}{\partial n} + h(\mathbf{r})\varphi(\mathbf{r}) = 0, \ \mathbf{r} \text{ on } S \tag{1.13}$$

where $h(\mathbf{r})$ is known function, $\frac{\partial \varphi(\mathbf{r})}{\partial n} = \mathbf{n}\nabla\varphi(\mathbf{r})$ is the directional derivative of φ along the outward normal to the boundary S, and $\mathbf{n}$ is a unit normal directed out of R. Note that the Neumann BC is a special case of the mixed condition with $h(\mathbf{r}) = 0$.

1.2   Maxwell Equations

The Maxwell equations are four differential equations which show classical properties of EM fields by using electric and magnetic fields. The equations are summarized in Table 1.1. Here, ρ_v is the electric volume charge density in C/m^3, $\mathbf{J}$ is the electric current density vector in A/m^2, $\mathbf{E}$ and $\mathbf{H}$ show the electric and magnetic field intensity vectors in V/m and A/m, respectively, and, $\mathbf{D}$ and $\mathbf{B}$ show the electric and magnetic flux density vectors in C/m^2 and Wb/m^2, respectively. The first two equations are related to the divergence of vectors, and the others are related to the curl operation of vectors. The relations between these vectors in simple medium are $\mathbf{D} = \varepsilon\mathbf{E}$, $\mathbf{B} = \mu\mathbf{H}$, $\mathbf{J} = \sigma\mathbf{E}$, where ε, μ, and σ denote the permittivity (F/m), the permeability (H/m), and the electric conductivity (S/m) of the medium.

Table 1.1   The Maxwell equations using differential form.

Name	Differential Form	Name	Differential Form
Gauss's law	$\nabla \cdot \mathbf{D} = \rho_v$	Gauss's law for magnetism	$\nabla \cdot \mathbf{B} = 0$
Faraday's law	$\nabla \times \mathbf{E} = -\dfrac{\partial \mathbf{B}}{\partial t}$	Maxwell–Ampere's law	$\nabla \times \mathbf{H} = \mathbf{J} + \dfrac{\partial \mathbf{D}}{\partial t}$

1.3 Guided Waves and Transverse/Longitudinal Decomposition

Guided wave propagation problems can be solved by using simplified equations obtained from the longitudinal and transverse decomposition of Maxwell equations that yield the transverse electric (TE), transverse magnetic (TM), and transverse electromagnetic (TEM) representations under different polarizations, such as perpendicular/parallel polarizations or horizontal/vertical polarizations in applications.

The open form of Maxwell curl operations of electric and magnetic fields for rectangular coordinate system can be written as

$$\left(\frac{\partial E_z}{\partial y} - \frac{\partial E_y}{\partial z}\right)\hat{\mathbf{x}} + \left(\frac{\partial E_x}{\partial z} - \frac{\partial E_z}{\partial x}\right)\hat{\mathbf{y}} + \left(\frac{\partial E_y}{\partial x} - \frac{\partial E_x}{\partial y}\right)\hat{\mathbf{z}}$$
$$= -\mu\frac{\partial H_x}{\partial t}\hat{\mathbf{x}} - \mu\frac{\partial H_y}{\partial t}\hat{\mathbf{y}} - \mu\frac{\partial H_z}{\partial t}\hat{\mathbf{z}} \qquad (1.14)$$

$$\left(\frac{\partial H_z}{\partial y} - \frac{\partial H_y}{\partial z}\right)\hat{\mathbf{x}} + \left(\frac{\partial H_x}{\partial z} - \frac{\partial H_z}{\partial x}\right)\hat{\mathbf{y}} + \left(\frac{\partial H_y}{\partial x} - \frac{\partial H_x}{\partial y}\right)\hat{\mathbf{z}}$$
$$= \left(J_x + \varepsilon\frac{\partial E_x}{\partial t}\right)\hat{\mathbf{x}} + \left(J_y + \varepsilon\frac{\partial E_y}{\partial t}\right)\hat{\mathbf{y}} + \left(J_z + \varepsilon\frac{\partial E_z}{\partial t}\right)\hat{\mathbf{z}}. \qquad (1.15)$$

A rectangular waveguide is a classical three-dimensional (3D) guiding structure. If a rectangular waveguide is located longitudinally along z-axis, TE/TM cases are defined by assuming no electric/magnetic field component in the direction of propagation, therefore the governing equations are given in Table 1.2.

The boundary should also be taken into consideration to determine the polarization type [1]. First of all, let us define the plane of incidence as the plane containing the normal to the boundary surface and the direction of propagation of the wave. For

Table 1.2   The governing equations for rectangular waveguide along z-axis.

TE Polarization	TM Polarization
$\dfrac{\partial E_y}{\partial z} = \mu \dfrac{\partial H_x}{\partial t}$	$\dfrac{\partial E_y}{\partial z} - \dfrac{\partial E_z}{\partial y} = \mu \dfrac{\partial H_x}{\partial t}$
$\dfrac{\partial E_x}{\partial z} = -\mu \dfrac{\partial H_y}{\partial t}$	$\dfrac{\partial E_z}{\partial x} - \dfrac{\partial E_x}{\partial z} = \mu \dfrac{\partial H_y}{\partial t}$
$\dfrac{\partial E_x}{\partial y} - \dfrac{\partial E_y}{\partial x} = \mu \dfrac{\partial H_z}{\partial t}$	$\dfrac{\partial E_y}{\partial x} = \dfrac{\partial E_x}{\partial y}$
$\dfrac{\partial H_z}{\partial y} - \dfrac{\partial H_y}{\partial z} = J_x + \varepsilon \dfrac{\partial E_x}{\partial t}$	$\dfrac{\partial H_y}{\partial z} = -J_x - \varepsilon \dfrac{\partial E_x}{\partial t}$
$\dfrac{\partial H_x}{\partial z} - \dfrac{\partial H_z}{\partial x} = J_y + \varepsilon \dfrac{\partial E_y}{\partial t}$	$\dfrac{\partial H_x}{\partial z} = J_y + \varepsilon \dfrac{\partial E_y}{\partial t}$
$\dfrac{\partial H_y}{\partial x} = \dfrac{\partial H_x}{\partial y}$	$\dfrac{\partial H_y}{\partial x} - \dfrac{\partial H_x}{\partial y} = J_z + \varepsilon \dfrac{\partial E_z}{\partial t}$

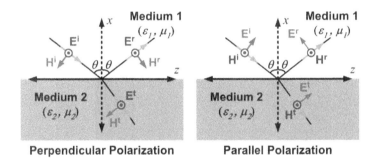

Perpendicular Polarization   **Parallel Polarization**

Figure 1.1   Perpendicular and parallel polarization on the zx-plane.

example, the plane of incidence is zx-plane in Fig. 1.1. Here, the electric field is either perpendicular to the plane of incidence for perpendicular polarization or parallel to the plane of incidence for parallel polarization. The governing equations are given in Table 1.3.

1.4   Two Dimensional Helmholtz's Equation

The wave equations for the electric and magnetic fields can be obtained by using Maxwell equations. If the electric and magnetic fields are to be time harmonic

Table 1.3 The governing equations for the plane of incidence on the zx-plane.

TE Polarization	TM Polarization
$\dfrac{\partial E_y}{\partial z} = \mu \dfrac{\partial H_x}{\partial t}$	$\dfrac{\partial H_y}{\partial x} = J_z + \varepsilon \dfrac{\partial E_z}{\partial t}$
$\dfrac{\partial E_y}{\partial x} = -\mu \dfrac{\partial H_z}{\partial t}$	$\dfrac{\partial H_y}{\partial z} = -J_x - \varepsilon \dfrac{\partial E_x}{\partial t}$
$\dfrac{\partial H_x}{\partial z} - \dfrac{\partial H_z}{\partial x} = J_y + \varepsilon \dfrac{\partial E_y}{\partial t}$	$\dfrac{\partial E_x}{\partial z} - \dfrac{\partial E_z}{\partial x} = -\mu \dfrac{\partial H_y}{\partial t}$

with the time dependence $\exp(-i\omega t)$, the wave equations in a linear, homogeneous, isotropic, source-free medium for each component of fields can be written as

$$\nabla^2 U - \mu\varepsilon \frac{\partial^2 U}{\partial t^2} = 0 \text{ (in time-domain)} \tag{1.16}$$

$$\nabla^2 U + k^2 U = 0 \text{ (in frequency-domain)} \tag{1.17}$$

where $k = \omega\sqrt{\mu\varepsilon}$ is the wavenumber, ω is the angular frequency, ∇^2 is the Laplace operator, and U shows the components of time harmonic either electric field or magnetic field. This is called homogeneous wave equation or Helmholtz's equation, that is the elliptic PDE. Assume an zx-plane as the two-dimensional (2D) environment, the Helmholtz's equation can be written as

$$\frac{\partial^2 U}{\partial z^2} + \frac{\partial^2 U}{\partial x^2} + k^2 U = 0. \tag{1.18}$$

1.5 Validation, Verification, and Calibration Procedure

Real-life engineering and EM problems can be handled via measurements or numerical simulations because only a limited number of problems with idealized geometries have mathematical exact solutions. The challenge in solving real-life engineering problems is therefore the reliability of the results. Reliability is achieved after a series of (model) validation, (data) verification, and (code) calibration (VV&C) tests [2].

Three fundamental building blocks of a simulation are the real-world problem entity being simulated, the conceptual model representation of that entity, and the computer implementation model. As illustrated in Fig. 1.2, engineers start with the definition of the real-life problem at hand. Electromagnetic problems, in general, are modeled with Maxwell equations and EM theory is well established by these equations. Maxwell equations are general and represent all linear EM problems. Once the geometry of the problem at hand (i.e., BCs) is given, they represent a unique

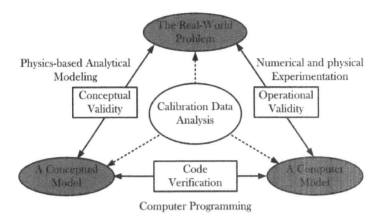

Figure 1.2   Validation, verification, and calibration procedure.

solution; the solution found by using Maxwell equations plus BC is the solution we are looking for. Unfortunately, there are only a few real-life problems which have mathematical exact solutions, therefore many different and approximate conceptual models can be used. It is the process of conceptual validity which shows that chosen conceptual model fits into the real-life problem the best under the specified initial and/or operational conditions. The next step is to develop a computer code for the chosen conceptual model. It is only after code verification via a computer programming process applied to show that the developed code represents the chosen conceptual model under given sets of conditions (accuracy, resolution, uncertainty, etc.). Finally, the solution for the real-life problem is obtained with confidence after numerical and/or physical experimentation; nothing but the operational validity process [3].

For the parabolic wave equation (PWE) chosen in this book, the VV&C procedure necessitates quantitatively and qualitatively answering these questions: (i) How precise is the PWE model? (ii) To what extent does the PWE correspond to the real-life problem? (iii) Under what/which conditions do different numerical methods yield reliable solutions? (iv) What is the accuracy of the numerical calculations?

1.6   Fourier Transform, DFT and FFT

The Fourier transform has been widely used in circuit analysis and synthesis, from filter design to signal processing, image reconstruction, etc. The reader should keep in mind that the time-domain and frequency-domain relations in electromagnetics are very similar to the relations between spatial and wavenumber domains. A simplest propagating (e.g., along z) plane wave is in the form of $\Phi(r, t) \propto e^{-i(\omega t - kz)}$ (where k and z are the wavenumber and position, respectively) and $\exp(-i\omega t)$ are also applicable to $\exp(ikz)$. Some characteristics are outlined as

- A rectangular time (frequency) window corresponds to a beam type (Sinc(·) function) variation in frequency (time)-domain.

- Similarly, a rectangular aperture (array) in spatial-domain corresponds to a beam type (Sinc(·) function) variation in wavenumber domain.

- The wider the antenna aperture the narrower the antenna beam; or, the narrower the pulse in time-domain the wider the frequency band.

Therefore, Fourier transform has also been used in electromagnetics from antenna analysis to imaging and non-destructive measurements, even in propagation problems. For example, the split-step parabolic equation (SSPE) method (which is nothing but the beam propagation method in optics) has been in use for several decades and is based on sequential Fourier transform operations between the spatial and wavenumber domains. Two- and three-dimensional propagation problems with non-flat realistic terrain profiles and inhomogeneous atmospheric variations above have been solved with this method successfully [3–5].

The principle of a transform in engineering is to find a different representation of a signal under investigation. The Fourier transform is the most important transform that is widely used in electrical engineering. The transformations between the time and the frequency-domains are based on the Fourier transform and its inverse Fourier transform. They are defined via

$$S(f) = \int_{-\infty}^{\infty} s(t) e^{i2\pi f t} dt, \text{ and } s(t) = \int_{-\infty}^{\infty} S(f) e^{-i2\pi f t} df. \tag{1.19}$$

Here, $s(t)$, $S(f)$, and f are the time signal, the frequency signal, and the frequency, respectively. We, the physicists and engineers, sometimes prefer to write the transform in terms of angular frequency $\omega = 2\pi f$, as

$$S(\omega) = \int_{-\infty}^{\infty} s(t) e^{i\omega t} dt, \text{ and } s(t) = \frac{1}{2\pi} \int_{-\infty}^{\infty} S(\omega) e^{-i\omega t} d\omega \tag{1.20}$$

which, however, destroys the symmetry. To restore the symmetry of the transforms, the convention is to divide $1/(2\pi)$ term into two and use $1/\sqrt{2\pi}$ during both Fourier transform and inverse Fourier transform. The Fourier transform is valid for real or complex signals, and, in general, is a complex function of ω (or f).

The Fourier transform is valid for both periodic and non-periodic time signals that satisfy certain conditions. Almost all real-world signals easily satisfy these requirements. It should be noted that the Fourier series is a special case of the Fourier transform. Mathematically, Fourier transform is defined for continuous time signals and in order to go to the frequency-domain, the time signal must be observed from an infinite-extend time window. Under these conditions, the Fourier transform defined above yields frequency behavior of a time signal at every frequency, with zero frequency resolution. Some functions and their Fourier transform are listed in Table 1.4. To compute the Fourier transform numerically on a computer, discretization plus numerical integration are required. This is an approximation of the true

Table 1.4   Some functions and their Fourier transforms.

Time Domain	Fourier Domain
Rectangular window	Sinc function
Sinc function	Rectangular window
Constant function	Dirac delta function
Dirac delta function	Constant function
Dirac comb (Dirac train)	Dirac comb (Dirac train)
Cosine function	Two real-even delta function
Sine function	Two imaginary-odd delta functions
Exponential function $\{\exp(-i\omega t)\}$	One positive-real delta functions
Gaussian function	Gaussian function

(i.e., mathematical), analytically defined Fourier transform in a synthetic (digital) environment, and is called the discrete Fourier transform (DFT). There are three difficulties with the numerical computation of the Fourier transform.

- *Discretization* (introduces periodicity in both time and frequency-domains),

- *Numerical integration* (introduces approximation and numerical round off and truncation errors),

- *Finite time duration* (introduces maximum frequency and resolution limitations).

The DFT of a continuous time signal sampled over a record period of T, with a sampling rate of Δt can be given as

$$S\left(m\Delta f\right) = \frac{T}{N} \sum_{n=0}^{N-1} s\left(n\Delta t\right) e^{i2\pi m\Delta f n\Delta t} \tag{1.21}$$

where $\Delta f = 1/T$, and, is valid at frequencies up to $f_{max} = 1/(2\Delta t)$. A simple MATLAB *dft_sin.m* file computes (1.21) for a time record $s(t)$ of two sinusoids whose frequencies are user specified. The record length and sampling time interval are also supplied by the user and DFT of this record is calculated inside a simple integration loop.

Let us plot two sinusoids with 10 Hz/1 V and 50 Hz/0.25 V both in time and frequency-domains. Choose $f_{max} = 200$ Hz, $\Delta f = 2$ Hz, $T = 0.5$ s, and $\Delta t = 2.5$ ms. Note that these parameters may be chosen arbitrarily in DFT but frequency resolution and maximum frequency will be $\Delta f = 1/T$ and $f_{max} = 1/(2\Delta t)$, respectively. Results are shown in Fig. 1.3.

The DFT requires an excessive amount of computation time, particularly when the number of samples N is high. The fast Fourier transform (FFT) is an algorithm to speed up DFT computations. The FFT forces one further assumption that N is an

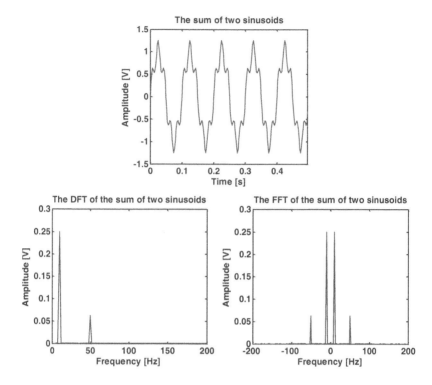

Figure 1.3 (Top) Time variations of two sinusoids, (bottom) frequency variations of two sinusoids obtained with (left) DFT and (right) FFT.

integer multiple of 2. This allows certain symmetries to occur reducing the number of calculations.

To write an FFT routine is not as simple as DFT routine, but there are many internet addresses where one can supply FFT subroutines (including source codes) in different programming languages, from Fortran to C++. Therefore, the reader does not need to go into details, rather include them in their codes by simply using *include* statements or *call* commands. In MATLAB, the calling command is *fft(s,N)* for the FFT and *ifft(S,N)* for the inverse FFT, where s and S are the recorded N-element time array and its Fourier transform, respectively. In order to do that one needs to replace the loop for DFT with a line code $Sf = fftshift(fft(st)*dt)$, that is included in *fft_sin.m* file. Note that one needs to scale the results in the frequency-domain (i.e., multiply the result by Δt) since MATLAB *fft(s,N)* command assumes $\Delta t = 1$; also, swap the first $N/2$ samples with the second half using the *fftshift(s)* command). Figure 1.3 also shows the FFT of the same signal.

As stated above, performing Fourier transform in a discrete environment introduces artificial effects. These are called aliasing effects, spectral leakage, and scalloping loss [3]. It should be kept in mind when dealing with DFT that

- Multiplication in the time-domain corresponds to a convolution in the frequency-domain.

- The Fourier transform of an impulse train in the time-domain is also an impulse train in the frequency-domain with the frequency samples separated by $T_0 = 1/f_0$.

- The narrower the distance between impulses (T_0) in the time-domain the wider the distance between impulses (f_0) in the frequency-domain (and vice versa).

- The sampling rate must be greater than twice the highest frequency of the time record, that is $\Delta t \geq 1/(2f_{max})$ (Nyquist sampling criterion).

- Since *time–bandwidth* product is constant, narrow transients in the time-domain possess wide bandwidths in the frequency-domain.

- In the limit, the frequency spectrum of an impulse is constant and covers the whole frequency-domain (that is why an impulse response of a system is enough to find out the response of any arbitrary input).

If the sampling rate in the time-domain is lower than the Nyquist rate, *aliasing* occurs [3]. Two signals are said to alias if the difference of their frequencies falls in the frequency range of interest, which is always generated in the process of sampling (aliasing is not always bad; it is called mixing or heterodyning in analog electronics and is commonly used in tuning radios and TV channels). It should be noted that although obeying Nyquist sampling criterion is sufficient to avoid aliasing, it does not give a high quality display in time-domain record. If a time signal sinusoid is not bin-centered in the frequency-domain then *spectral leakage* occurs. In addition, there is a reduction in coherent gain if the frequency of the sinusoid differs in value from the frequency samples, which is termed *scalloping loss*.

Fourier transform is used for *energy signal* which contains finite energy. This means $\int_{-\infty}^{\infty} |s(t)|^2 \, dt$ is finite. Periodic functions do not satisfy this property. *Power signals* have finite power in one period (i.e., $\frac{1}{P} \int_{-P/2}^{P/2} |s(t)|^2 \, dt$ is finite where P is the period of the signal). Power signals are represented in terms of Fourier series. A function $f(x)$ is periodic, with period P, if

$$f(x) = \sum_{n=1}^{\infty} f(x + nP). \tag{1.22}$$

A periodic function $f(x)$ can be approximated by using Fourier series expansion as

$$f(x) \approx \frac{A_0}{2} + \sum_{n=1}^{\infty} A_n \cos\left(\frac{2\pi nx}{P}\right) + B_n \sin\left(\frac{2\pi nx}{P}\right) \tag{1.23}$$

where

$$A_0 = \frac{2}{P} \int_{-P/2}^{P/2} f(x) \, dx \tag{1.24}$$

$$A_n = \frac{2}{P} \int_{-P/2}^{P/2} f(x) \cos\left(\frac{2\pi nx}{P}\right) dx \qquad (1.25)$$

$$B_n = \frac{2}{P} \int_{-P/2}^{P/2} f(x) \sin\left(\frac{2\pi nx}{P}\right) dx. \qquad (1.26)$$

Non-periodic functions may also be approximated by Fourier series inside a finite region by assuming the finite region as the period of that function. In this case, it should be remembered that the Fourier series representation no longer represents the function outside the region. Equations (1.24)–(1.26) show that, one needs to multiply the function with sine and cosine functions along the whole period and then integrate in order to find out Fourier coefficients A_n and B_n.

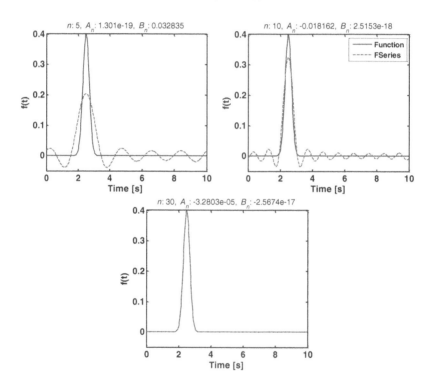

Figure 1.4   Gaussian function and its Fourier series approximation with 5, 10, 30 terms.

Fseries.m lists a simple MATLAB code for the Fourier series representation of a given function. A few examples plotted with this MATLAB code are given in Fig. 1.4.

The Gaussian function used in this example is

$$f(t) = \frac{\exp\left(-12.5(t - 2.5)^2\right)}{\sqrt{2\pi}} \qquad (1.27)$$

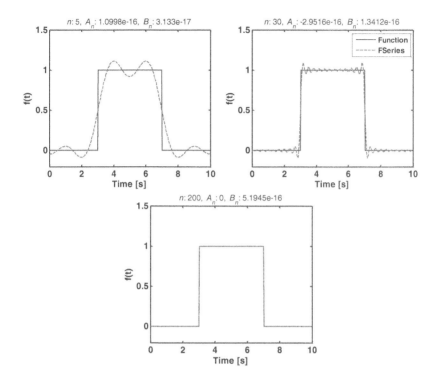

Figure 1.5   Rectangular pulse and its Fourier series approximation with 5, 30, 200 terms.

and the interval (i.e., the period) is $0 \leq t \leq 10$. First, only the first five terms are used and the result is plotted in Fig. 1.4. The solid and dashed lines in the figures correspond to the function and its Fourier series approximation, respectively. Second figure corresponds to the same scenario but with the first ten terms in the series summation. Finally, the last figure belongs to the same comparisons with the first thirty terms. The agreement in curves in Fig. 1.4 shows that, thirty terms are adequate for this function in this interval (period).

As shown above, any piecewise continuous function may be approximated by a series summation of sine and cosine functions. The number of terms required in the Fourier series representation depends on the smoothness of the function and the specified accuracy. The degree of smoothness of the function determines the number of terms in its Fourier representation. In addition, only sine or cosine terms contribute the function if it is odd or even symmetric. Understanding digital communication concepts, the frequency content of the rectangular pulse should be well analyzed. A symmetric rectangular pulse is defined as

$$\text{Rect}\left(\frac{t}{T}\right) = \begin{cases} 1 & -\frac{T}{2} \leq t \leq \frac{T}{2} \\ 0 & t > \left|\frac{T}{2}\right| \end{cases} \tag{1.28}$$

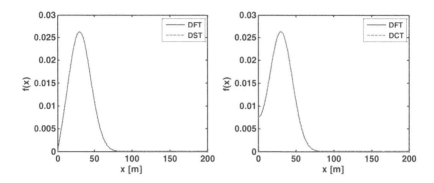

Figure 1.6 Comparison of DFT and DST/DCT analysis for the given Gaussian antenna profile: (left) DBC, (right) NBC.

and an infinite number of terms is required to fully represent this function with the Fourier series summation. The terms are called harmonics. It is interesting to visualize term by term contributions in the Fourier series representation. This is illustrated in Fig. 1.5 for a 4 s.-rectangular pulse between 3 s. and 7 s.

Fourier transform and Fourier series expansion are important procedures in engineering [6]. The DFT and FFT are discrete tools to analyze time-domain signals. One needs to know the problems caused because of the discretization and specify the parameters accordingly to avoid non-physical and non-mathematical results. Moreover, extra attention should be paid when using built-in commands in different computer languages (e.g., MATLAB). As pointed out above in the text, one needs to multiply the results of the FFT taken using MATLAB's *fft(s,N)* command with Δt in order to obtain correct amplitude values.

Since the DFT cannot handle the BCs automatically in propagation problems, the discrete sine transform (DST) and discrete cosine transform (DCT) can be used for various BCs on Earth. Using DFT, to satisfy the BC over perfect electric conductor (PEC) ground, the boundary is extended from [0, hmax] to [-hmax, hmax], and then, in accordance with the image theory, the odd and even symmetric field profiles are constructed for Dirichlet boundary condition (DBC) and Neumann boundary condition (NBC), respectively, to be able to apply the DFT. Another option, to avoid the height extension, is to reduce DFT to one-sided DST or DCT, for DBC and NBC, respectively. A MATLAB code *fft_dst_dct.m* compares DFT and DST/DCT of a Gaussian field profile. A 30 m height antenna with 0.1° beamwidth at 3 GHz is used in Fig. 1.6.

In addition, MATLAB functions for Fourier transforms are given below. Note that the initial and end values of *s* are zero for DST.

DFT of field	*fftshift(fft(ifftshift(s)))*	Inverse DFT of field	*fftshift(ifft(ifftshift(S)))*
DST of field	*dst(s(2:end-1))*	Inverse DST of field	*[0;idst(S);0]*
DCT of field	*dct(s)*	Inverse DCT of field	*idct(S)*

CHAPTER 2

WAVE PROPAGATION OVER FLAT EARTH

2.1 Flat Earth and GO Two-Ray Model

The simplest propagation scenario used in analytical modeling is the flat Earth with PEC surface under line-source excitation. Figure 2.1 shows the propagation environment over PEC flat Earth. The total field at the observer is obtained via vector addition of direct ray and ground-reflected ray. This model is called geometric optics (GO) two-ray (2Ray) model [7]. Rays in two dimensions are attenuated by the square root of the distance as they propagate away from the source, therefore the total field is

$$u\left(z_o, x_o\right) = H_0^{(1)}\left(kd_1\right) + \Gamma H_0^{(1)}\left(kd_2\right) \tag{2.1}$$

where z and x are the longitudinal and vertical coordinates, respectively; $H_0^{(1)}$ is the first-kind Hankel function with order zero, k is the wavenumber, Γ is the reflection coefficient, d_1 is the distance between the source at (z_s, x_s) and the observer at (z_o, x_o) (direct ray), d_2 is the distance between image source at $(z_s, -x_s)$ and the observer (ground-reflected ray), with

$$d_1 = \sqrt{d^2 + (x_s - x_o)^2}, \; d_2 = \sqrt{d^2 + (x_s + x_o)^2} \tag{2.2}$$

Radio Wave Propagation and Parabolic Equation Modeling, First Edition. By Gökhan Apaydin, Levent Sevgi
© 2017 by the Institute of Electrical and Electronic Engineers, Inc. Published 2017 by John Wiley & Sons, Inc.

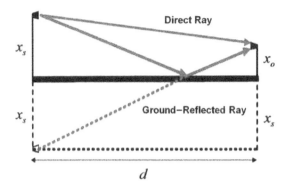

Figure 2.1 Construction of two rays from the source to the observer.

and d is the longitudinal distance between the source and the observer.

The reflection coefficient becomes $\Gamma = -1$ and $\Gamma = 1$ for the horizontal and vertical polarizations, respectively, assuming PEC surface. $u(z, x)$ represents the electric and magnetic fields for the horizontal and vertical polarizations, respectively. Here the electric/magnetic field has only one nonzero component E_y/H_y for horizontal/vertical polarization, respectively. The radiation condition applies along $x \rightarrow \infty$, $z \rightarrow \pm\infty$ and both the DBC and NBC are taken into consideration at the PEC surface for horizontal and vertical polarizations, respectively.

If the observer is far away from the source, the exponential functions may also be used instead of Hankel functions as

$$H_0^{(1)}(kd) \cong \sqrt{\frac{-2i}{\pi}} \frac{e^{ikd}}{\sqrt{kd}}. \tag{2.3}$$

Ray2.m is a short MATLAB code for 2Ray model. First, fields versus range/height variations are mapped in color plots. Figure 2.2 shows 3D color maps for horizontal and vertical polarizations, respectively. In this example, the frequency is 30 MHz. The source is located 200 m above the ground. The longitudinal and transverse mesh sizes are $\Delta z = 100$ m and $\Delta x = 1$ m, respectively.

Then, horizontal propagation factor (PF) at 100 m height and vertical PF at 300 m and 600 m ranges are computed using *Ray2PF.m* in Fig. 2.2. The frequency is 30 MHz as above, but the source is located 150 m above the ground with vertical polarization. The longitudinal and transverse mesh sizes of this example are $\Delta z = 1$ m and $\Delta x = 0.1$ m, respectively.

2.2 Single Knife Edge Problem and Four-Ray Model

On the single knife edge problem, the ray summation approach is based on the construction of four different rays, related reflection, and diffraction coefficients using Fresnel integrals [8]. The single knife edge problem, the four-ray (4Ray) model,

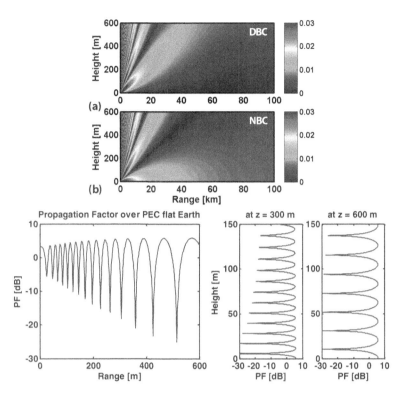

Figure 2.2   (Top) Fields vs. range/height under line-source excitation over PEC flat Earth: (a) horizontal polarization, DBC, (b) vertical polarization, NBC. (Bottom) PF (left) at 100 m height, (right) at 300 m and 600 m ranges.

and Fresnel integral representations can be used as an alternative model from which reliable reference data can be generated. The scenario of this canonical problem is pictured in Fig. 2.3. Here, h_t, h_r, and h_w are the heights of the transmitter, receiver, and the knife edge obstacle; d_1 and d_2 are the distances from the source to the obstacle and from the obstacle to the receiver, respectively.

Possible four rays are as follows: Ray 1 is the direct path between the transmitter and the receiver. Ray 2 is considered as the ray from the transmitter reflected from the right side of the knife edge obstacle. This ray reaches the receiver directly or tip diffraction may occur. Ray 3 is considered as the ray from the transmitter reflected from the left side of the knife edge obstacle. Same as before, this ray also reaches the receiver directly or tip diffraction will occur. Ray 4 is considered as the ray from the transmitter reflected from both the left side and the right side of the knife edge obstacle.

The parameters of the Fresnel integrals are derived by using the image source and receiver for the reflected waves. The Fresnel clearance, the height of the knife edge above the line of sight (LOS) may be positive or negative [8]. When the direct ray

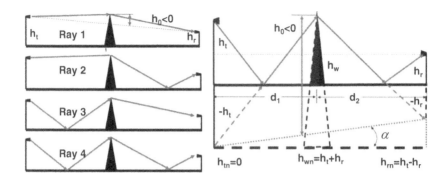

Figure 2.3   Geometry of flat Earth for single knife edge: (left) possible four rays, (right) construction of Ray 4 from source/receiver images.

between the transmitter and receiver intersects obstacle, h_0 is taken as negative. The Fresnel integrals $C(v)$ and $S(v)$ are evaluated, where $v = h_0\sqrt{2}$ with h_0 equal to the ray clearance over the knife edge. The propagation factor (PF) is equal to

$$PF = \frac{E}{E_0} = \sum_{q=1}^{4} E_q \exp(i\psi_q) \tag{2.4}$$

where $E_q = A(v_q)\Gamma_q$, $\Gamma_1 = 1$, $\Gamma_2 = \Gamma_R$, $\Gamma_3 = \Gamma_L$, $\Gamma_4 = \Gamma_L\Gamma_R$, and $\psi_q = \beta_q + (R_q - R_1)k$. Here,

$$A(v_q) = \sqrt{\frac{(C(v_q) + 0.5)^2 + (S(v_q) + 0.5)^2}{2}} \tag{2.5}$$

for $C(v_q) = \int_0^v \cos(v_q^2)dv$, $S(v_q) = \int_0^v \sin(v_q^2)dv$; the distances of four rays are

$$R_q = \begin{cases} \sqrt{d^2 + (h_t - h_r)^2} & \text{for } q = 1, q = 3 \\ \sqrt{d^2 + (h_t + h_r)^2} & \text{for } q = 2, q = 4 \end{cases} \tag{2.6}$$

$$\beta_q = \begin{cases} \arctan\left(\frac{S(v_q)+0.5}{C(v_q)+0.5}\right) - \frac{\pi}{4} & \text{if } C(v_q) \geq -0.5 \\ \arctan\left(\frac{S(v_q)+0.5}{C(v_q)+0.5}\right) + \frac{3\pi}{4} & \text{if } C(v_q) < -0.5 \end{cases} \tag{2.7}$$

and the complex reflection coefficients, for the horizontal and vertical polarizations, respectively, are

$$\Gamma_h = \frac{\sin\theta - \sqrt{\varepsilon - \cos^2\theta}}{\sin\theta + \sqrt{\varepsilon - \cos^2\theta}}, \; \Gamma_v = \frac{\varepsilon\sin\theta - \sqrt{\varepsilon - \cos^2\theta}}{\varepsilon\sin\theta + \sqrt{\varepsilon - \cos^2\theta}} \tag{2.8}$$

where $\varepsilon = \varepsilon_r + i60\sigma_g\lambda$, σ_g is the conductivity of ground, and ε_r is the relative permittivity of ground, θ is the angle of incidence. The ray clearances for the four rays are

$$h_{01} = \sqrt{\frac{2d}{\lambda d_1 d_2}} \left(-h_t + \frac{(h_r - h_t)d_1}{d} - h_w \right) \tag{2.9}$$

$$h_{02} = \sqrt{\frac{2d}{\lambda d_1 d_2}} \left(-h_t + \frac{(h_r + h_t)d_1}{d} - h_w \right) \tag{2.10}$$

$$h_{03} = \sqrt{\frac{2d}{\lambda d_1 d_2}} \left(h_t + \frac{(-h_r - h_t)d_1}{d} - h_w \right) \tag{2.11}$$

$$h_{04} = \sqrt{\frac{2d}{\lambda d_1 d_2}} \left(-h_t + \frac{(h_t - h_r)d_1}{d} - h_w \right). \tag{2.12}$$

Using MATLAB code *Ray4.m*, Fig. 2.4 shows 3D color maps for horizontal and vertical polarizations, respectively. The scenario belongs to one-way propagation over PEC ground with 75 m height-wall at 15 km range. The line source is at 15 m height at $z = 0$. The frequency is 3 GHz. The longitudinal and transverse mesh sizes are $\Delta z = 100$ m and $\Delta x = 0.1$ m, respectively. The PFs versus height in front of and beyond the obstacle are also plotted in Fig. 2.4. Here, two vertical field profiles are obtained with the 4Ray model. The first plot on the left belongs to the interference region (before the obstacle); the other one is in the diffraction region (beyond the obstacle).

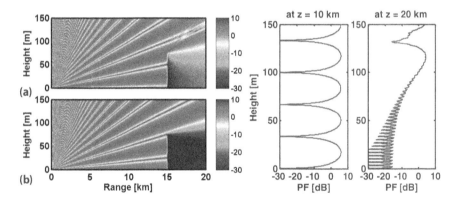

Figure 2.4 (Left) PF vs. range/height using 4Ray model over PEC flat Earth with single knife edge: (a) horizontal polarization, DBC, (b) vertical polarization, NBC. (Right) PF vs. height at 10 km and 20 km ranges, horizontal polarization ($f = 3$ GHz, source height is 15 m, $\Delta z = 100$ m, $\Delta x = 0.1$ m, obstacle is 75 m height-wall at 15 km range).

2.3   Vertical Linear Refractivity Profile and Mode Summation

Propagation with a linearly decreasing vertical refractivity profile is

$$n^2(x) = 1 - a_0 x \tag{2.13}$$

where $a_0 > 0$ is a refractivity parameter that controls the strength of the duct and the PEC flat Earth is a canonical structure to generate the reference solutions [7]. The 2D propagation scenario is completed by choosing the appropriate transverse and longitudinal BC. This problem has analytical exact solutions in terms of Airy functions for the range independent vertical refractive index. The modal series solution can be expressed as

$$u(z,x) = \sum_{q=1}^{\infty} c_q \psi_q(x) e^{i\beta_q z} \tag{2.14}$$

where c_q are the modal excitation coefficients, β_q is the longitudinal propagation constant for the related mode represented by index q. The new function $\psi_q(x)$ satisfies the one-dimensional (1D) wave equation in the transverse-domain

$$\left[\frac{d^2}{dx^2} + k_0^2 n^2(x) - \beta_q^2 \right] \psi_q(x) = 0. \tag{2.15}$$

Using $k_0^2 n^2(x) - \beta_q^2 = K_1 x + K_2$ and $\rho = -K_1^{-2/3}(K_1 x + K_2)$, the new variables are obtained for a linearly decreasing vertical refractivity in (2.13) as $K_1 = -a_0 k_0^2$, $K_2 = k_0^2 - \beta_q^2$, $\rho = (a_0 k_0^2)^{1/3} x - (a_0 k_0^2)^{-2/3}(k_0^2 - \beta_q^2)$; and the wave equation becomes the Airy equation [8]

$$\left[\frac{d^2}{d\rho^2} - \rho \right] \psi_q(\rho) = 0. \tag{2.16}$$

Then, the exact modal solution of Airy equation in (2.16) can be written as [7]

$$u(z,x) = \sum_{q=1}^{N} c_q Ai \left[(a_0 k_0^2)^{1/3} x - \sigma_q \right] e^{i\beta_q z} \tag{2.17}$$

where Ai is the first-kind of Airy function and the longitudinal propagation constant for the related mode represented by index q is

$$\beta_q = \pm\sqrt{k_0^2 - (a_0 k_0^2)^{2/3} \sigma_q}. \tag{2.18}$$

The BC at the surface is satisfied with $Ai(-\sigma_q) = 0$, $Ai'(-\sigma_q) = 0$ for the DBC and NBC, respectively. Here, the prime denotes the derivative with respect to the vertical coordinate. The problem is then reduced to finding the modal excitation coefficients from a given antenna pattern using orthonormality property from a given source function as

$$c_q = \int_0^{X_{max}} g(x) Ai \left[(a_0 k_0^2)^{1/3} x - \sigma_q \right] dx \tag{2.19}$$

where

$$g(x) = \frac{1}{\sqrt{2\pi\sigma^2}} \exp\left[\frac{(x - x_s)^2}{2\sigma^2} \right]. \tag{2.20}$$

Here, σ is the spatial width and x_s is the height of the Gaussian source $g(x)$. The Gaussian source pattern is often used in applications since it represents various antenna types (but any other source profile may also be used). The Gaussian antenna pattern can also be defined in the vertical wavenumber domain as

$$g(k_x) = \exp\left[\frac{-k_x^2 \ln 2}{2k_0^2 \sin^2(\theta_{bw}/2)}\right]. \tag{2.21}$$

The tilt (or elevation) angle (θ_{elv}) is introduced by shifting the antenna pattern, that is $g(k_x) \to g(k_x - k_0 \sin\theta_{elv})$. The vertical field in the spatial-domain is then obtained by taking the inverse Fourier transform of (2.21).

The fundamental issue here is the construction of the reference data. An antenna radiation pattern may be used for transmitter modeling which is mathematically achieved by locating a vertical Gaussian pattern, $g(x) = u(z_s, x)$, on a specified height. Then, the modal summation in (2.17) is used together with the orthonormality condition (2.19) and the number of modes and their excitation coefficients are derived numerically for a given error boundary.

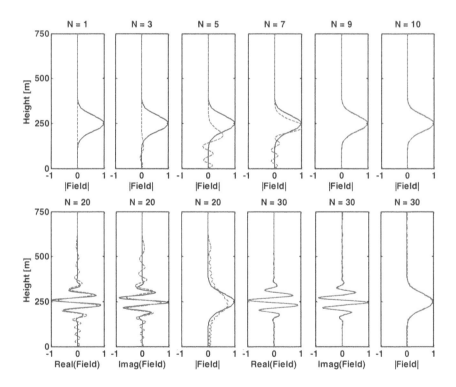

Figure 2.5   Contribution of modes to a given (top) untilted and (bottom) tilted $1°$ upward Gaussian antenna patterns under linearly decreasing refractivity profile (horizontal polarization, $f = 300$ MHz, $x_s = 250$ m, $\theta_{bw} = 0.35°$, $a_0 = 1.2 \times 10^{-6}$).

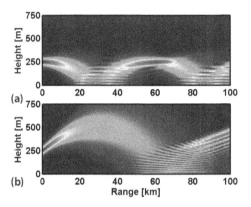

Figure 2.6   Three-dimensional color plots of electric fields vs. range–height variations for a given Gaussian antenna pattern (a) untilted, (b) tilted $1°$ upward using mode summation (horizontal polarization, $f = 300$ MHz, $\theta_{bw} = 0.5°$, $a_0 = 1.2 \times 10^{-6}$, $N_x = 256$, $N_z = 1000$).

Finally, vertical boundaries of the numerical integration during modal excitation coefficient extraction from the orthonormality property increase as the mode number increases. The specification of the number of numerical integration steps for the calculation of modal excitation coefficients is crucial. The code must adopt the number of integration steps automatically as the mode index increases.

A novel MATLAB code *Linear.m* is developed for the calculation of mode summation over with a linearly decreasing vertical refractivity profile.

Figure 2.5 shows the contribution of modes for untilted and tilted Gaussian antenna patterns, respectively. Note that modal excitation coefficients are real if the antenna pattern has no vertical tilt (i.e., the antenna pattern is horizontal, parallel to the flat Earth). These modal excitation coefficients become complex when upslope or downslope tilt is introduced. Moreover, the modes are confined between the Earth's surface and modal caustics which depend on the mode number; the higher the mode, the higher the location of the caustic. Therefore, the number of modes used in the superposition directly depends on the antenna height.

In Fig. 2.6, the 2D propagation space and propagation over the non-penetrable flat Earth is simulated with and without antenna tilt. The antenna has a sectoral vertical radiation beam. The vertical, linearly decreasing refractivity profile causes a surface duct as observed in the figure. Although the beam is directed horizontally (Fig. 2.6a) and upward (Fig. 2.6b), EM waves are pushed down as they propagate longitudinally, hit the ground, and reflect upward. The ground-reflected waves and the interference (between the incident and the ground-reflected waves) are clearly observed in this figure.

CHAPTER 3

PARABOLIC EQUATION MODELING

3.1 Introduction

Radio engineers or site surveyors dream to have numerical propagation tools that calculate the groundwave path loss (PL) between any two points marked on their digital maps. This necessitates the solution of the EM wave equation in three dimensions in a manner that takes into account various EM effects, such as, the irregular terrain profile, the vegetation, the Earth's curvature, the atmospheric refractivity, the presence of buildings, cars, and other obstacles. Also, the solution must include all the relevant scattering components (e.g., multiple reflections and refractions, edge/tip diffractions, surface and/or leaky waves) that account for the PL [3, 4, 9]. Unfortunately, this has been not yet in sight.

The current numerical strategy is to take the 2D projections between the transmitter and the receiver and reduce the 3D problem in hand into 2D range–height scenarios. Highly attractive 2D propagation prediction virtual tools have been developed for the last decade. These virtual tools are based on various techniques, such as the PE [10], the method of moments (MoM) [11], PE and MoM [12], the finite-difference time domain (FDTD) [13], the transmission line matrix (TLM) [14], ray-mode summation [15, 16], and ray tracing, ray shooting [17]. Among these, the

Radio Wave Propagation and Parabolic Equation Modeling, First Edition. By Gökhan Apaydin, Levent Sevgi
© 2017 by the Institute of Electrical and Electronic Engineers, Inc. Published 2017 by John Wiley & Sons, Inc.

PE-based propagation tools are speculated to be the most attractive and effective ones.

Two of these virtual tools (i.e., [10] and [12]) are of special interest in terms of the topic discussed in this area. The SSPE_GUI presented in [10] is a simple MATLAB-based groundwave propagation package for the visualization of EM propagation over irregular terrain through the non-homogeneous atmosphere, for waves radiated by a horizontally oriented antenna over the ground. A MATLAB package which modifies the MoM by the application of the forward/backward spectral acceleration technique [11] and integrates it with the SSPE method is introduced in [12]. With this, precise SSPE vs. MoM comparisons are possible.

The PE was first applied to the underwater acoustic propagation modeling by Leontovich and Fock in the 1940s [18]. One major problem at that time was to detect the nuclear submarines as early as possible, and this certainly necessitated a good understanding of underwater acoustic propagation in deep as well as shallow waters with variable densities, bottom profiles, streams, etc. Later, it has become popular in EM wave propagation modeling through the Earth–troposphere waveguides. Although Leontovich and Fock [18, 19] were the pioneers, the PE approach becomes famous after the introduction of the Fourier SSPE algorithm [20]. Since then, the PE technique has been improved, combined with many auxiliary tools, and applied to a variety of complex propagation problems. The book by Levy [5] is a good source which discusses the PE modeling in detail and gathers a huge list of PE related studies. There are also several surveys on PE modeling such as the one in [21].

The finite-element method (FEM) has also been used in developing PE-based numerical propagation tools. Initially, the FEM-based PE models were also applied in underwater acoustic propagation prediction problems (see [22]). A few FEM-based PE models have also appeared in EM wave propagation modeling for the last decade [23, 24].

3.2   Parabolic Wave Equation Form

The problem of PWE is the 2D EM problem bounded by the Earth's surface at the bottom, and unbounded at the top, that is extending to infinity. The problem is governed by the scalar Helmholtz's equation

$$\frac{\partial^2 U}{\partial z^2} + \frac{\partial^2 U}{\partial x^2} + k_0^2 n^2 U = 0 \tag{3.1}$$

where $k_0 = 2\pi/\lambda$ is the free-space wavenumber (λ is the wavelength), $n = n(z, x)$ is the refractive index, z and x stand for the longitudinal (range) and transverse (height above ground) coordinates, respectively. Furthermore, $U(z, x)$ corresponds to either electric or magnetic field components for horizontal and vertical polarizations. Note that some researchers use the terms perpendicular and parallel polarizations, whereas others prefer horizontal or vertical polarizations. Here, we assume zx-plane as the 2D environment. The TM (vertical polarization) and TE (horizontal polarization) equations use (E_x, H_y, E_z) and (H_x, E_y, H_z), respectively [1]. Therefore, U is H_y and E_y for TM and TE polarizations, respectively.

By separating the rapidly varying phase term $\exp(ik_0 z)$ from the field $U(z, x) = \exp(ik_0 z)u(z, x)$ in (3.1) if the direction of wave propagation is predominantly along the $+z$-axis (i.e., paraxial direction) the following two equations can be obtained:

$$\left(\frac{\partial^2}{\partial z^2} + \frac{\partial^2}{\partial x^2} + k_0^2 n^2 \right) \exp(ik_0 z)u(z, x) = 0 \tag{3.2}$$

$$\exp(ik_0 z) \left(\frac{\partial^2}{\partial z^2} + 2ik_0 \frac{\partial}{\partial z} + \frac{\partial^2}{\partial x^2} + k_0^2(n^2 - 1) \right) u = 0. \tag{3.3}$$

Since $\exp(ik_0 z)$ is not always zero for all z values, the PWE in terms of slowly varying amplitude function $u(z, x)$ in range reduces as follows:

$$\left(\frac{\partial^2}{\partial z^2} + 2ik_0 \frac{\partial}{\partial z} + \frac{\partial^2}{\partial x^2} + k_0^2(n^2 - 1) \right) u = 0 \tag{3.4}$$

$$\left(\frac{\partial}{\partial z} + ik_0(1 - Q) \right) \left(\frac{\partial}{\partial z} + ik_0(1 + Q) \right) u = 0 \tag{3.5}$$

where $Q = (1 + q)^{1/2}$ and $q = k_0^{-2} \partial^2 / \partial x^2 + (n^2 - 1)$ if the refractive index is range independent. Note that this assumption does not violate the applicability of PWE to range dependent refractivity profiles because it is valid for each range step during the split-step solution of the PWE, which will be clear in the sequel. The first and second parts of (3.5) correspond to the forward- and backward-propagating waves, respectively. If the backward propagation is ignored, (3.5) reduces to

$$\left(\frac{\partial}{\partial z} + ik_0(1 - Q) \right) u = 0. \tag{3.6}$$

The formal solution of the forward propagation part in (3.6) can be expressed as

$$u(z + \Delta z, x) = \exp\left(-ik_0 \Delta z(1 - Q) \right) u(z, x) \tag{3.7}$$

which is amenable to numerical solution by marching-type algorithms along the range. For the square root approximation of

$$(1 + q)^{1/2} \approx \frac{a_0 + a_1 q}{b_0 + b_1 q} \tag{3.8}$$

(3.6) can be written as

$$\left\{ \begin{aligned} & \left(b_0 + b_1 \left(n^2 - 1 \right) \right) \frac{\partial}{\partial z} + \left(b_1 k_0^{-2} \right) \frac{\partial^3}{\partial x^2 \partial z} \\ & + ik_0 \left((b_0 - a_0) + (b_1 - a_1)(n^2 - 1) \right) + ik_0^{-1}(b_1 - a_1) \frac{\partial^2}{\partial x^2} \end{aligned} \right\} u = 0. \tag{3.9}$$

Equation (3.9) can be generally considered

$$\left(A_0 \frac{\partial}{\partial z} + A_1 \frac{\partial^3}{\partial x^2 \partial z} + A_2 + A_3 \frac{\partial^2}{\partial x^2} \right) u = 0 \tag{3.10}$$

where $A_0 = b_0 + b_1(n^2 - 1)$, $A_1 = b_1 k_0^{-2}$, $A_2 = ik_0\left((b_0 - a_0) + (b_1 - a_1)(n^2 - 1)\right)$, $A_3 = ik_0^{-1}(b_1 - a_1)$. If the angle of propagation measured from paraxial direction is less than $15°$, the standard PE is obtained with the help of first-order Taylor expansion (square root approximation) $(1 + q)^{1/2} \approx 1 + q/2$; therefore $a_0 = 1$, $a_1 = 0.5$, $b_0 = 1$, $b_1 = 0$ and $A_0 = 1$, $A_1 = 0$, $A_2 = -0.5ik_0(n^2 - 1)$, and $A_3 = -0.5ik_0^{-1}$,

$$\left\{\frac{\partial}{\partial z} - 0.5ik_0(n^2 - 1) - 0.5ik_0^{-1}\frac{\partial^2}{\partial x^2}\right\} u = 0. \qquad (3.11)$$

By multiplying $2ik_0$, the standard PE is derived as

$$\frac{\partial^2 u}{\partial x^2} + 2ik_0\frac{\partial u}{\partial z} + k_0^2(n^2 - 1)u = 0. \qquad (3.12)$$

The accuracy of the standard PE can be limited to propagation angles less than $15°$, and the error in the approximation increases with $\sin^4 \theta$, where θ is the propagation angle from the horizontal, due to the first neglected term in Taylor's expansion. Hence, the standard PE is known as the narrow-angle approximation to the wave equation. Since the propagation angles encountered in long-range propagation problems are usually less than a few degrees, the accuracy of the standard PE is adequate for numerical modeling.

In problems involving large propagation angles more than $15°$ (such as short-range propagation problems or the problems exhibiting strong multipath effects), a more accurate expansion of the operator Q is required. In such cases, the use of higher-order polynomials for the operator causes instabilities in the numerical results. There are various convenient methods proposed in the literature to handle large propagation angles such that the operator Q can be written as $Q = \sqrt{1 + A + B}$, and approximated as $Q \approx \sqrt{1 + A} + \sqrt{1 + B} - 1$ where $A = \partial^2/k_0^2\partial x^2$ and $B = n^2 - 1$. By making use of the operator identity $\sqrt{1 + A} = 1 + A\left(\sqrt{1 + A} + 1\right)^{-1}$, the wide-angle PE is given by [25, 26]

$$\frac{\partial u}{\partial z} - \left[ik_0^{-1}\left(\sqrt{1 + \frac{1}{k_0^2}\frac{\partial^2}{\partial x^2}} + 1\right)^{-1} + ik_0(n - 1)\right] u = 0. \qquad (3.13)$$

Claerbout equation can be also obtained by using the first-order Pade approximation $(1 + q)^{1/2} \approx (1 + 0.75q)/(1 + 0.25q)$; therefore $a_0 = 1$, $a_1 = 0.75$, $b_0 = 1$, $b_1 = 0.25$ and $A_0 = 1 + 0.25(n^2 - 1)$, $A_1 = 0.25k_0^{-2}$, $A_2 = -0.5ik_0(n^2 - 1)$, $A_3 = -0.5ik_0^{-1}$ in (3.10) to satisfy the propagation angles up to $40°$ [27] as

$$\left\{k_0^2(n^2 + 3)\frac{\partial}{\partial z} + \frac{\partial^3}{\partial x^2\partial z} - 2ik_0^3(n^2 - 1) - 2ik_0\frac{\partial^2}{\partial x^2}\right\} u = 0. \qquad (3.14)$$

Another alternative for wide-angle cases is to use Greene approximation to satisfy propagation angles up to $40°$–$45°$, $\left(\sqrt{1 + q} \approx (0.99987 + 0.79624q)/(1 + 0.30102q)\right)$, therefore $a_0 = 0.99987$, $a_1 = 0.79624$, $b_0 = 1$, $b_1 = 0.30102$ and $A_0 = 1 +$

$0.30102(n^2 - 1)$, $A_1 = 0.30102k_0^{-2}$, $A_2 = ik_0 \left(0.00013 - 0.49522(n^2 - 1)\right)$, $A_3 = -0.49522ik_0^{-1}$ as [28]

$$\left\{\begin{array}{l} \left(1 + 0.30102 \left(n^2 - 1\right)\right) \dfrac{\partial}{\partial z} + 0.30102k_0^{-2} \dfrac{\partial^3}{\partial x^2 \partial z} \\ +ik_0 \left(0.00013e - 0.49522(n^2 - 1)\right) - 0.49522ik_0^{-1} \dfrac{\partial^2}{\partial x^2} \end{array}\right\} u = 0. \quad (3.15)$$

3.3   Dirichlet, Neumann, and Cauchy Boundary Conditions

The 2D propagation scenario is completed by choosing the appropriate transverse and longitudinal BCs. Figure 3.1 shows the 2D projected propagation region. Here, x and z are the height and range coordinates, respectively. The region is N-segments vertically and M-segments longitudinally. The transverse BC over Earth's surface is expressed as

$$\left(\alpha_1 \frac{\partial}{\partial x} + \alpha_2\right) u(z, x) = 0 \quad (3.16)$$

where α_1 and α_2 are constants. In general, an impedance boundary condition (IBC) or CBC (i.e., lossy Earth's surface) is introduced via $\alpha_1 = 1$, $\alpha_2 = ik_0\sqrt{\varepsilon - 1}$ and $\alpha_1 = 1$, $\alpha_2 = ik_0\sqrt{\varepsilon - 1}/\varepsilon$ for horizontal and vertical polarization, respectively. Here, $\varepsilon = \varepsilon_r + i60\sigma_g\lambda$ is the complex relative dielectric constant in terms of relative permittivity (ε_r) and conductivity (σ_g) of the lossy ground at range z [29].

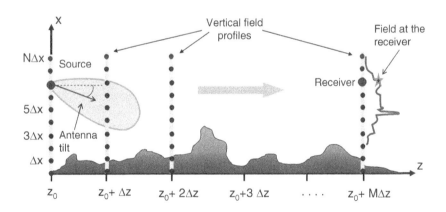

Figure 3.1   The 2D projected propagation region.

For the PEC surface ($\sigma \to \infty$), impedance BC is replaced with either DBC or NBC, corresponding to the horizontal and vertical polarizations, respectively. The cases where $\alpha_1 = 0$ and $\alpha_2 = 0$ refer to DBC and NBC, respectively, over the PEC surface.

Since the propagation problem involves a physical domain extending vertically to the infinity, an abrupt truncation is required at certain height, and therefore, the upper

BC (in other words, free-space or open-region condition) must be satisfied to avoid non-physical reflections. Such artificial reflections can be obviated by introducing absorbing layers above the height of interest [5].

Since waves propagating upward either go to infinity or bent down because of the refractivity variations, a window function (i.e., Hanning, Hamming, or Tukey) can be applied to the vertical field profile above the selected height at each range in order to eliminate reflection effects.

The radiation BC along z is

$$\left\{ \frac{\partial}{\partial z} - ik_0 \right\} u(z,x) \Bigg|_{z \to \pm\infty} \to 0. \tag{3.17}$$

Although the PE in 2D describes one-way propagation and cannot take the backscattered waves into account, this is not a serious restriction in order to investigate waves emanating from a transmitter and reaching a receiver. It should be noted that the solution of (3.10) in 2D yields waves that are attenuated by the square root of the distance as they propagate away from the transmitter (i.e., cylindrical waves spread for an infinite-length line source which allows the reduction of the 3D wave equation into 2D). Therefore, the user should divide the field strength results by the square root of distance in order to obtain PL vs. range variations in a realistic 3D environment [30].

3.4 Antenna/Source Injection

In most of the 2D propagation models, the source is modeled in a transmitting antenna's pattern that is located vertically at the initial range. This can be achieved by using a vertical Gaussian field profile.

The initial vertical field at the starting range position must be properly determined in accordance with the parameters of the antenna's pattern being modeled. The initial field can be computed by means of near-field/far-field transformation that relates the aperture field and beam pattern, along with the utilization of FFT. The height and elevation angle of the antenna can be included by using Fourier shift theorems. The antenna pattern $g(k_x)$ given in (2.21) is specified by three parameters: antenna height x_s, the 3 dB beamwidth θ_{bw}, and the tilt (or elevation) angle θ_{elv}. The first step is to specify Gaussian antenna pattern in the k_x-domain via

$$\tilde{u}(0, k_x) = g(k_x)\exp(-ik_x x_s) + \Gamma g^*(-k_x)\exp(ik_x x_s). \tag{3.18}$$

The tilt is introduced by replacing $g(k_x)$ with $g(k_x - k_0 \sin\theta_{elv})$. The initial field profile is then obtained via inverse FFT of $\tilde{u}(0, k_x)$. The initial field in the spatial-domain can also be represented as $u(0,x) = u_s(0,x) + \Gamma u_s(0,-x)$ using

$$u_s(0,x) = \exp\left(ik_0 x \sin\theta_{elv} - \frac{(x - x_s)^2}{w^2} \right) \tag{3.19}$$

where $w = \sqrt{2\ln 2}/(k_0 \sin(\theta_{bw}/2))$. The Gaussian antenna pattern is often used in applications since it represents various antenna types (such as parabolic antennas).

3.5 Split-Step Parabolic Equation (SSPE) Model

The standard PE given in (3.12) turns out to be a first-order ordinary differential equation that can be solved directly if a Fourier transform from the x domain to the spectral k_x domain is applied. Although, in real problems, n can be a function of height and/or range (i.e., $n = n(z, x)$), this approach is appropriate because the equation is solved at each small-range step size, Δz, which is chosen small enough so that within any Δz interval the refractive index can be assumed to be constant with respect to z. The numerical SSPE solution for $j = 1, 2, ..., M$ is given as [31]

$$
\begin{aligned}
u(z_s + j\Delta z, x) = \quad & \exp\left[i\tfrac{k_0}{2}(n^2 - 1)\Delta z\right] \\
& \times F^{-1}\left\{\exp\left[-i\tfrac{k_x^2 \Delta z}{2k_0}\right] F\left\{u(z_s + (j-1)\Delta z, x)\right\}\right\}.
\end{aligned} \quad (3.20)
$$

This equation can be used to calculate $u(z, x)$ along z with the steps of Δz, once the initial field distribution, $u(z_s, x)$, is given. A 1D array can be used to store the transverse field profiles at N vertical height points at M discrete ranges with replacement. The standard SSPE procedure is as follows:

- An antenna pattern representing the initial height profile of the field, $u(z_s, x)$, is injected first. The initial profile is propagated longitudinally from z_s to $z_s + \Delta z$ for $j = 1$ via (3.20) and $u(z_s + \Delta z, x)$ is obtained. This new height profile is then used as the initial profile for the next step and $u(z_s + 2\Delta z, x)$ is obtained for $j = 2$ via (3.20). The procedure is applied repeatedly, and the vertical field profiles are stored at each range step until the desired range.

- The SSPE sequentially operates between the x and k_x-domains, which are the Fourier transform pairs. Since the transforms are implemented numerically, the domains are truncated at $\pm X_{max}$ and $\pm k_{xmax}$. The transverse and longitudinal step sizes Δx and Δz, respectively, and the maximum height, X_{max}, are determined from the source and/or observation requirements, as well as sampling necessities and aliasing effects. Once X_{max} is determined, k_{xmax} is found from the Nyquist sampling criteria, $X_{max} \times k_{xmax} = \pi L$, with L being the transform size. Note that $k_{xmax} = k_0 \sin\theta_{max}$, where θ_{max} is the maximum allowable propagation angle. Since $\Delta x = X_{max}/L$, the altitude increment should satisfy $\Delta x \leq \lambda/(2\sin\theta_{max})$. Although the choice of Δx is quite critical in simulations, the selection of the range increment Δz (taking the refractivity gradients into account) is chosen by the user and can be much larger than the wavelength.

- The SSPE cannot automatically handle the BCs at the surface. It is satisfied at the surface by extending the initial vertical profile from $[0 - X_{max}]$ to $[-X_{max}, X_{max}]$ (odd and even symmetric for DBC and NBC, respectively).

- The problem at hand has a vertically semi-open propagation region ($x \to \infty$); therefore, an abrupt truncation is required at a certain height, which means that, strong artificial reflections will occur if not taken care of. These non-physical reflections can be removed by using artificial absorbing layers above the height of interest. This is achieved by adding a small imaginary part to the refractive index (by making n complex) or by applying windowing functions (such as Hanning, Hamming, Tukey) in order to eliminate reflection effects.

- Irregular terrain is implemented within the SSPE algorithm via a terrain file and a staircase approximation [10].

- The PEC ground modeling is adequate for many applications (especially above 100–200 MHz), but a more accurate model is required below these frequencies. The surface impedance implementation (i.e., CBC) involves additional modifications but is not too difficult. The discrete mixed Fourier transform (DMFT) extends the SSPE algorithm to account for the ground losses [32–34].

In the FFT-based SSPE algorithm, to satisfy the BCs over the Earth's surface, the surface is removed by taking a mirror copy of the initial vertical field profile with respect to the surface (odd/even symmetric for DBC/NBC, respectively). Another choice is to use one-sided sine/cosine transforms (sinFFT/cosFFT) for DBC/NBC, respectively.

3.5.1 Narrow-Angle and Wide-Angle SSPE

The numerical solution of the PE is achieved by the Fourier SSPE method, which is a widely used and robust algorithm. The algorithm starts at a reference range (usually at an antenna) and marches the solution in a range in such a way that, it obtains the vertical field profile at a given range by using the field at the previous range, with appropriate BCs at the top and bottom boundaries of the domain. The split-step solution of the narrow-angle PE in (3.12) is given by

$$
\begin{aligned}
u(z + \Delta z, x) = \quad & \exp\left[ik_0(n^2 - 1)\tfrac{\Delta z}{2}\right] \\
& \times F^{-1}\left\{\exp\left[-ik_x^2 \tfrac{\Delta z}{2k_0}\right] F\left\{u(z, x)\right\}\right\}
\end{aligned}
\tag{3.21}
$$

where F indicates the Fourier transform and $k_x = k_0 \sin\theta$ is the transform variable (θ is the propagation angle from the horizontal).

The wide-angle split-step solution of (3.12) is given as [25, 26]

$$
\begin{aligned}
u(z + \Delta z, x) = \quad & \exp\left[ik_0(n - 1)\Delta z\right] \\
& \times F^{-1}\left\{\exp\left[-ik_x^2 \frac{\Delta z}{k_0 + \sqrt{k_0^2 - k_x^2}}\right] F\left\{u(z, x)\right\}\right\}.
\end{aligned}
\tag{3.22}
$$

3.5.2 A MATLAB-Based Simple SSPE Code

A short MATLAB code is given below step by step with explanations in order to understand SSPE model easily. This code calculates 3D color plots of field vs. range–height variations at 300 MHz. A given Gaussian antenna pattern of beamwidth $\theta_{bw} = 1°$ is tilted $\theta_{tilt} = 1°$ upward at 100 m height under horizontal polarization. The refractivity parameter given in (2.13) is chosen $a_0 = 2.4 \times 10^{-6}$ for the ducting case. Figure 3.2 shows the results.

```
% WAVE PROPAGATION ANALYSIS OVER FLAT EARTH USING SSPE
clc, close all, clear all
%% STEP 1: Input parameters are specified
zmax = 50; xmax = 500; % Maximum range [km], Maximum height [m]
NZ = 501;  NX = 1001;  % Number of nodes on z and x dir.
% Gaussian antenna specifications
fr = 300; xs = 100;% Frequency [MHz], Antenna height [m]
bw = 1;   elv = 1; % Beamwidth angle [deg], Elevation angle [deg]
pol = 1;     % Polarization type: 1 for 'Horz.' or 0 for 'Vert.'
rf = [1.5 2];% filter coefficients applied to [rf1 rf2]*xmax
mdl = 'Narr';% 'Narr' or 'Wide' for Narrow—angle/Wide—angle SSPE
a0 = 2.4e—6; % refractivity parameter for n^2=1—a0x use Eq.(2.13)
```

```
%% STEP 2: Initialize parameters for SSPE analysis
% Using (2.13) refractive index squared function(n^2) is defined.
REFRSQ=@(x) 1—a0*x;
% Filter function is defined to eliminate reflection
FILFUN=@(x) (x<=rf(1)*xmax)+(x>rf(1)*xmax).*...
    (1+cos(pi*(x—rf(1)*xmax)/((rf(2)—rf(1))*xmax)))/2;
% Using (3.19) Gaussian source function is defined.
k0=2*pi*fr*1e6/3e8; % wavenumber
ww=sqrt(2*log(2))/(k0*sind(bw/2));
ufs=@(x) exp(1i*k0*sind(elv)*x).*exp(—((x—xs)/ww).^2);
ufss=@(x) ufs(x)+(—1)^pol*ufs(—x);
% The nodal points and increments (dz,dx) are defined.
dz=1e3*zmax/(NZ—1); z=(0:dz:1e3*zmax).';
dx=xmax/(NX—1);     x=(—rf(2)*xmax:dx:rf(2)*xmax).';
% The initial field is obtained.
Field=ufss(x); % initial field
NXm=length(x); % number of nodes between —rf2*hmax and rf2*hmax
ind0hmax=find(x>=0&x<=xmax); % nodes between 0 and xmax
% filter function is obtained.
flt=FILFUN(abs(x));
% SSPE parameters are obtained.
kx=2*pi/(dx*NXm)*(—ceil((NXm—1)/2):NXm—1—ceil((NXm—1)/2));
if strcmp(mdl,'Narr'),
    % Narrow—angle SSPE using Eq.(3.21)
    CC1=exp(—1i*dz*kx.^2/(2*k0)).';
    CC2=exp( 1i*dz*k0*(REFRSQ(abs(x))—1)/2);
elseif strcmp(mdl,'Wide'),
    % Wide—angle SSPE using Eq.(3.22)
    CC1=exp(—1i*dz*kx.^2./(sqrt(k0^2—kx.^2)+k0)).';
    CC2=exp( 1i*dz*k0*(sqrt(REFRSQ(abs(x)))—1));
end
```

```
%% STEP 3: Start SSPE analysis
u=zeros(NX,NZ); u(:,1)=Field(ind0hmax);
for n=2:NZ,
    % FFT of u(z,:) is calculated and multipled by CC1
    Field=CC1.*fftshift(fft(ifftshift(Field),NXm));
    % inv.FFT is calculated,multiplied by CC2 and filter function
    Field=CC2.*fftshift(ifft(ifftshift(Field),NXm)).*flt;
    u(:,n)=Field(ind0hmax);% new fields are injected to u(z+dz,:)
end
```

```
%% STEP 4: Show 3D fields
x=x(ind0hmax); u=abs(u); u=u/max(u(:,1));
f=figure('Position',[20 80 700 300]);
axes('Parent',f,'LineWidth',2,'FontWeight','bold','FontSize',14);
hold on,box on,surf(z/1000,x,u);shading interp;view([0 90]),
ylim([0 xmax]);cmax=max(u(:));caxis([0 cmax/1.4]),
colorbar('FontWeight','bold','FontSize',14);
xlabel('Range [km]'); ylabel('Height [m]');
```

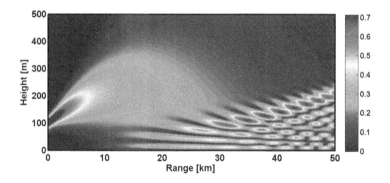

Figure 3.2 The result of MATLAB code.

3.6 FEM-Based Parabolic Equation Model

The initial phase of the FEM-based PE (FEMPE) procedure is to divide the transverse domain between the ground and the user-defined maximum height (X_{max}) into a number of elements. Next, starting from the initial field at $z = z_s$, the approximated field values at the selected discrete nodes in the vertical domain are propagated longitudinally by applying the Crank–Nicolson approach. This is based on the improved Euler method [35–37]. Although the Crank–Nicolson technique is inherently fast, it requires that, both height and range step sizes, Δx and Δz, respectively, should be chosen as small as necessary to overcome numerical oscillation problems, which obviously decelerate the speed of the method.

Table 3.1 The coefficients of (3.10) with respect to narrow and wide cases.

Coefficients	Narrow	Wide Pade(1,1)	Wide (Greene)
A_0	$2ik_0$	$k_0^2(n^2 + 3)$	$1 + 0.30102(n^2 - 1)$
A_1	0	1	$0.30102k_0^{-2}$
A_2	$k_0^2(n^2 - 1)$	$-2ik_0^3(n^2 - 1)$	$ik_0(0.00013 - 0.49522(n^2 - 1))$
A_3	1	$-2ik_0$	$-0.49522ik_0^{-1}$

To formulate the FEMPE, we start with (3.10) whose coefficients are chosen by considering narrow- and wide-angle approximations given in Table 3.1. Note that some coefficients depend on x because of the refractive index, therefore the midpoints of each segment have been taken into consideration for the matrix calculation. However, the coefficients are assumed to be constant for each element by finding the value of the refractive index at the midpoint of each element. Multiplying (3.10) by a smooth test function while considering the BC; integrating from $x = 0$ to $x = X_{max}$, it is obtained as [35]

$$\sum_{e=1}^{n_e} \left(\sum_{j=1}^{2} A_0 \frac{\partial c_j^e}{\partial z} \int_{x_1^e}^{x_2^e} B_j^e B_m^e \, dx - \sum_{j=1}^{2} A_1 \frac{\partial c_j^e}{\partial z} \int_{x_1^e}^{x_2^e} \frac{\partial B_j^e}{\partial x} \frac{\partial B_m^e}{\partial x} \, dx \right.$$
$$\left. + \sum_{j=1}^{2} A_2 c_j^e \int_{x_1^e}^{x_2^e} B_j^e B_m^e \, dx - \sum_{j=1}^{2} A_3 c_j^e \int_{x_1^e}^{x_2^e} \frac{\partial B_j^e}{\partial x} \frac{\partial B_m^e}{\partial x} \, dx \right) \quad (3.23)$$

$$+ A_3 \alpha_2 \delta_{m,1} c_1^1 = 0$$

with Kronecker delta ($\delta_{m,\widetilde{m}} = 0$ for $m \neq \widetilde{m}$, $\delta_{m,\widetilde{m}} = 1$ for $m = \widetilde{m}$). The idea of the FEM is to divide the domain, here the vertical profile $[0 - X_{max}]$, into subdomains (called elements) in which the basis functions are generally formed with the help of linear piecewise Lagrange polynomials as [35]

$$B_1^e(x) = \frac{x_2^e - x}{x_2^e - x_1^e}, \quad B_2^e(x) = \frac{x - x_1^e}{x_2^e - x_1^e} \quad (3.24)$$

where e stands for the elements between nodes x_1^e and x_2^e. The matrix form can then be represented as

$$\mathbf{A} \frac{\partial \mathbf{c}}{\partial \mathbf{z}} + \mathbf{B} \mathbf{c} = \mathbf{0} \quad (3.25)$$

with $\mathbf{A} = A_0\mathbf{M} - A_1\mathbf{K}$ and $\mathbf{B} = A_2\mathbf{M} - A_3\mathbf{K} + A_3\mathbf{BC}$ or
$\left[A_0 M_{mj}^e - A_1 K_{mj}^e \right] \left\{ \frac{\partial c_j^e}{\partial z} \right\} + \left[A_2 M_{mj}^e - A_3 K_{mj}^e + A_3 BC_{mj}^e \right] \left\{ c_j^e \right\} = \{0\}$ for

$e = 1, ..., n_e$, $m = 1, 2$, and $j = 1, 2$ with

$$M_{mj}^e = \int_{x_1^e}^{x_2^e} B_m^e B_j^e dx, \ K_{mj}^e = \int_{x_1^e}^{x_2^e} \frac{\partial B_m^e}{\partial x} \frac{\partial B_j^e}{\partial x} dx. \tag{3.26}$$

The elemental matrices between nodes x_1^e and x_2^e for the linear piecewise Lagrange polynomials are obtained as

$$[M^e] = \frac{\Delta x}{6} \begin{bmatrix} 2 & 1 \\ 1 & 2 \end{bmatrix}, \ [K^e] = \frac{1}{\Delta x} \begin{bmatrix} 1 & -1 \\ -1 & 1 \end{bmatrix}. \tag{3.27}$$

BC_{mj}^e is taken into consideration with CBC as $BC_{11}^1 = \alpha_2$ for each step in z, $z + \Delta z, ...$ The ground properties are incorporated into the FEM with the help of matrix **BC**. On the PEC surface, matrix **BC** becomes zero. In addition, the initial node is always zero for DBC at the surface and this is satisfied by eliminating the first column and row of the matrices.

Using Crank–Nicolson approximation based on the improved Euler method for $k = 2, ..., N_z$ [29] where M is the number of nodes in the horizontal domain, the new coefficients for the next step are obtained as

$$\mathbf{c}^k = \mathbf{c}^{k-1} + \frac{\Delta z}{2} \left(\frac{\partial \mathbf{c}^k}{\partial z} + \frac{\partial \mathbf{c}^{k-1}}{\partial z} \right) \tag{3.28}$$

multiplying by **A**, and eliminating the derivative terms by using the differential equation (3.25), one obtains

$$\left(\mathbf{A} + \mathbf{B} \frac{\Delta z}{2} \right) \mathbf{c}^k = \left(\mathbf{A} - \mathbf{B} \frac{\Delta z}{2} \right) \mathbf{c}^{k-1} \tag{3.29}$$

which yields an unconditionally stable system and accurate method with a discretization error $O\left(\Delta z^2\right)$. The coefficients of the initial field $\mathbf{c}^1$ at $z = 0$ can be generated from the Gaussian antenna pattern specified by its height, 3 dB beamwidth angle, and tilt angle. Although Crank–Nicolson provides a fast solution, it has some disadvantages since oscillation occurs for large Δz [35].

For example, for the linearly decreasing vertical refractivity (see Section 2.3 for details) problem, the LINPE package is created to compare SSPE and FEMPE with analytical mode summation representation. The flow chart of the LINPE is given in Fig. 3.3. The propagation tool has three subroutines (LIN_EXACT, SSPE, and FEMPE) for the numerical calculations of the mathematical exact, the DFT/DST/DCT-based PE, and the FEMPE solutions, respectively. MATLAB codes, *sspe.m* and *fempe.m*, are used for calculations. First, the 2D propagation environment (the maximum range and height) is specified, the frequency is given, and the discretization parameters (Δx, Δz, the number of range $N_z = M + 1$, and height points $N_x = N + 1$, etc.) are supplied. Then, the Gaussian antenna pattern (i.e., the antenna height, its beamwidth, and vertical tilt) is given as the initial vertical field profile and then normalized to unity for all simulation tests. Finally, the slope of the vertically decreasing refractivity profile is supplied.

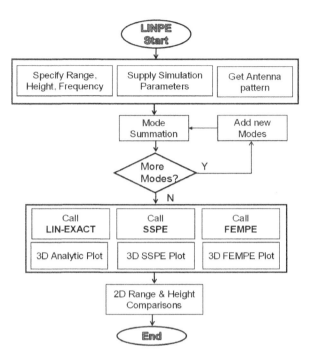

Figure 3.3   The flow chart of the MATLAB-based LINPE code. LINPE predicts the field strength value at a given range and height point using all three methods; analytical exact representation, SSPE, and FEMPE.

LINPE propagation tool links to the mode summation subroutine once all operational parameters are set. The mode subroutine calculates the number of modes and the modal excitation coefficients that are necessary to construct the initial antenna pattern for the given accuracy. These modes are then used for every height and range points to calculate the mathematical solution that will be used as the reference.

LINPE calls all three subroutines one by one, does the numerical computations and yields the output as the 2D and 3D field strength variations. It yields not only 3D field strength vs. range–height color plots for all the three methods but also the 2D and plots for their comparisons, field strength vs. height variations are plotted at three different ranges. Field strength vs. range variations is given at two different heights.

The tests and comparisons belong to the scenario specified with the operational parameters listed in Table 3.2.

Figure 3.4 presents 3D color plots of fields vs. range–height for a given Gaussian antenna pattern without a tilt via the LIN_EXACT, SSPE, and FEMPE subroutines for horizontal and vertical polarizations. The vertical, linearly decreasing refractivity profile causes a surface duct as observed in the figure. Although the beam is

Table 3.2 The operational parameters.

Operational Parameter	Value	Operational Parameter	Value
Maximum range [km]	100	Number of range [pts]	1000
Maximum height [m]	750	Number of height [pts]	256
Frequency [MHz]	300	Antenna height [m]	250
Vertical beamwidth [°]	0.5	Antenna tilt [°]	0/1
Refractivity slope (a_0)	1.2×10^{-6}		

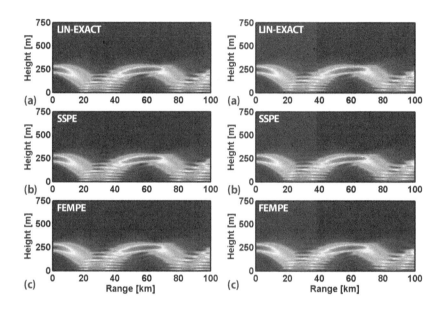

Figure 3.4 Three-dimensional color plots of fields vs. range–height variations for a given Gaussian antenna pattern without a tilt obtained from (a) analytical, (b) SSPE, (c) FEMPE representations: (left) horizontal polarization, (right) vertical polarization.

directed horizontally, EM waves are pushed down as they propagate longitudinally, hit the ground, and reflect upward. The ground-reflected waves and the interference (between the incident and the ground-reflected waves) are clearly observed in this figure.

Unfortunately, these 3D color plots are mostly good for the visualization purposes and are not good for the realistic comparisons and the calibration. The calibration of such numerical propagation tools necessitates accurate comparisons. This is achieved via two dimension range and/or height illustrations of EM fields as presented in Fig. 3.5. Here, the vertical field variations computed via all three methods at three different ranges are given for different polarizations. On the left, the initial field profile (i.e., the antenna radiation pattern) is shown. The other two show the vertical field profiles at 25 km and 50 km ranges, respectively.

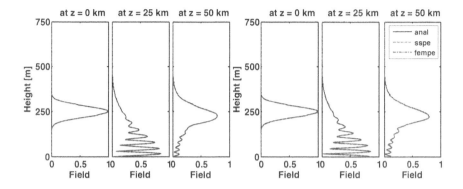

Figure 3.5   Fields vs. height at three specified ranges; (solid) analytical representations, (dashed) SSPE, (dashed-dotted) FEMPE, (left) horizontal polarization, (right) vertical polarization.

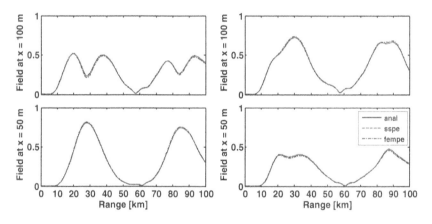

Figure 3.6   Fields vs. range at two specified heights; (solid) analytical representations, (dashed) SSPE, (dashed-dotted) FEMPE, (left) horizontal polarization, (right) vertical polarization.

Figure 3.6 shows the same comparison for different polarizations among these three methods as field strengths vs. ranges at two different heights, 50 m and 100 m, respectively.

In the second test, the 2D propagation space and propagation over the flat non-penetrable ground is simulated with an antenna tilt. Different linearly decreasing refractivity profiles are used. Although the beam is directed upward, EM waves are pushed down again as they propagate longitudinally, hit the ground, and reflect up-ward in Fig. 3.7. The ground-reflected waves and the interference (between the inci-dent and the ground-reflected waves) are clearly observed in this figure. As observed

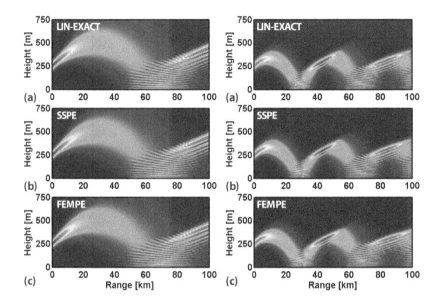

Figure 3.7   Three-dimensional color plots of fields vs. range–height variations for a given Gaussian antenna pattern tilted $1°$ upward obtained from (a) analytical, (b) SSPE, (c) FEMPE representations, horizontal polarization, (left) $a_0 = 1.2 \times 10^{-6}$, (right) $a_0 = 3.6 \times 10^{-6}$.

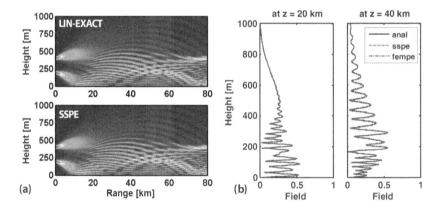

Figure 3.8   (a) SSPE and analytic propagators with tilted waves at 200 m and 400 m with $-0.5°$, $0.5°$ tilts, respectively, (b) vertical field profiles at two different ranges ($a_0 = 1.2 \times 10^{-6}$).

from these three figures, the agreement between the SSPE and FEMPE numerical propagation tools and their match with the analytical exact solutions are impressive.

The analytical calculations are much faster if the antenna beamwidth is wide (i.e., the number of modes is less). The narrower the vertical antenna pattern the higher the number of modes used to reconstruct this pattern, therefore the longer the computation time. Note that the tilt of the antenna in analytical exact representations is satisfied by introducing the complex modal excitation coefficients.

The tests performed here and comparisons against the analytical exact solution which may be accepted as reference calibrate both SSPE and FEMPE propagation tools.

In Fig. 3.8a, 3D visualization of both analytical and numerical solutions is presented where the transmitter contains two Gaussian patterns of 1° beamwidth (i.e., two antennas) at 200 m and 400 m, with 0.5° downward and 0.5° upward, respectively. In Fig. 3.8b, vertical field profiles at two different ranges obtained with all three (analytical, SSPE, and FEMPE) codes are shown. Excellent agreement illustrates the success and completeness of numerical methods.

Important aspects of these tests, comparisons, and calibration may be listed as follows:

- The modes of the analytical exact solution are confined between the ground and their (modal) caustics for the linearly decreasing vertical refractivity profile. The height of the caustics increases as the number of modes increases, therefore the construction of the initial field profile necessitates the contributions of higher-order modes for higher antenna heights. But, as the antenna height increases, the contributions of the lower-order modes become negligible, so the best way to reconstruct the antenna pattern in terms of the modal summation is to use modes from M_1 to M_2 which are the lowest and the highest mode numbers to be determined from operational parameters.

- The modal excitation coefficients are found from the given antenna pattern using the orthonormality property. These coefficients are real for the antenna pattern without tilt but become complex if a tilt is used. The complex excitation coefficients are obtained from a numerical complex integration.

- An antenna pattern may be reconstructed with a number of modes for a given set of operational parameters. But, one should take higher-order modes into account (although they might not be necessary to construct antenna pattern) if the antenna is tilted upward.

- The analytical solution is exact, therefore is a *reference* when an infinite number of modes is taken into account and if accurately computed numerically. On the other hand, the analytical solutions with a finite number of modes should be accurate enough to be a *reference*!

- SSPE solves the PE in the transverse-domain so it is not as sensitive as FEMPE to the vertical discretization. On the other hand, special attention is required for the vertical boundary treatments.

- FEMPE uses discretization on the vertical-domain and solves the discrete Crank–Nicolson system horizontally. The computation burden mostly comes from the selection of this tridiagonal system. The computation time increases with N_z in range but N_x^2 in height.

The aim in this chapter is to (i) review both the SSPE and FEMPE propagation approaches and introduce MATLAB-based simple routines; (ii) present canonical test cases and do systematic comparisons between the SSPE and FEMPE methods; (iii) calibrate both the SSPE and FEMPE tools against analytical exact (reference) solutions. The canonical tests and comparisons are given. Both the SSPE and FEMPE tools are also calibrated against the analytical exact solutions.

The PE method is capable of simulating one-way, forward-scatter propagation problems within the paraxial approximation. It takes into account all types of BCs at the ground, as well as the atmospheric refractivity variations. Widely applied numerical routines are based on either the discrete Fourier split-step solution or the FEM solution. Here, two numerical propagation tools based on the Fourier split-step and the FEM are discussed. Simple MATLAB-based propagation tools (LIN_EXACT, SSPE, and FEMPE) are introduced. The systematic comparisons on some canonical test scenarios are performed and the propagation tools are calibrated against the mathematical exact solutions. Both SSPE and FEMPE can take any type of refractivity variations into account.

3.7   Atmospheric Refractivity Effects

The lower troposphere affects radio wave propagation in numerous ways. Especially, its non-uniform nature causes EM waves to be bent or refracted. Refraction of EM waves is due to the variation of the velocity of propagation with altitude. It is known that index of refraction or refraction index (n) is the ratio of the velocity in free space to the velocity in the medium of interest (atmosphere in our case), and is caused by pressure, temperature, and water vapor variations in both space and time. The air can be considered as a non-dispersive medium and represented by its refractive index ($n = \sqrt{\varepsilon_r}$). The refractivity of the propagation medium should be well understood since non-flat terrain and/or Earth's curvature can be implemented via refractivity in most of the analytical and numerical approaches. Refractive index of the air is very close to unity (typically about 1.0003), therefore it is customary to use refractivity

$$N = (n - 1) \times 10^6 \tag{3.30}$$

where N is dimensionless but is measured in N units for convenience. The refractivity depends on the pressure P (mbar), the absolute temperature T (°K), and the partial pressure of water vapor e (mbar)

$$N = 77.6 \frac{P}{T} + 3.73 \times 10^5 \frac{e}{T^2} \tag{3.31}$$

which is valid in Earth–troposphere waveguides and can be used in radio wave propagation modeling. If the refractive index is constant, radio waves would propagate in straight lines. Since n decreases with height, radio waves are bent downward toward the Earth, so that the radio horizon lies further away than the optical horizon. It should be noted that the radio horizon effect is taken into account either by using N with the effective Earth radius or by introducing a fictitious medium where N is replaced by the modified refractivity M, which is also dimensionless but is measured in M units

$$M = N + \frac{x}{a_e} \times 10^6 = N + 157x \qquad (3.32)$$

with the height above surface (x) given in km and the effective Earth's radius, that is, $a_e = 6,378$ km. It is conventional to define modified refractivity, which takes the Earth's curvature into account, as follows

$$M = (n + x/a_e - 1) \times 10^6. \qquad (3.33)$$

The variations of the vertical gradient dM/dx of the modified refractivity determine four types of atmospheric conditions.

- Sub-refraction ($dM/dx >$117 M units/km),

- Standard ($dM/dx =$ 117 M units/km),

- Super-refraction ($dM/dx <$117 M units/km), and

- Ducting ($dM/dx <$0).

Non-standard atmospheric conditions cause anomalous propagation because rays bend upward in sub-refraction, and downward to the Earth's surface in super-refraction and ducting conditions, in a way different from the standard atmosphere.

Especially, atmospheric ducts (i.e., wave trapping layers) are of special interest because the negative vertical gradient leads to the capture of energy within the duct, and the trapped energy can propagate to ranges beyond the normal horizon what would be expected with a standard atmosphere. Such conditions significantly affect the radio communication links and radar performance. There are basically four types of ducting conditions

- surface duct,

- surface-based duct,

- elevated duct,

- evaporation duct (which is indeed a type of surface duct occurring over water due to water vapor evaporated from the sea).

For the standard atmosphere (i.e., for a vertical linearly decreasing refractive index), N decreases by about 40 N units/km while M increases by about 117 N units/km. Sub-refraction (super-refraction) occurs when the rate of change in N with respect to height (i.e., dN/dx) is less (more) than 40 N units/km.

A linearly decreasing (increasing) vertical refractive index variation forces a wave trap near (diverge from) the Earth's surface while propagating. Similar effects are also caused by concave and convex surfaces. Therefore, there is an analogy between refractive index and surface geometry in terms of propagation effects. By using this analogy, the Earth's curvature effect is included into the refractive index of the air. The Earth's curvature effect is equivalent to a vertical refractivity gradient of 157 N units/km (i.e., linear vertical increasing refractivity profile). As stated, the standard atmosphere, together with the Earth's curvature, is represented by a vertical refractivity gradient of 117 N units/km.

CHAPTER 4

WAVE PROPAGATION AT SHORT RANGES

4.1 Introduction

Propagation above the Earth has long been modeled in two dimensions, where ground losses and irregularities and atmospheric variations have been taken into account, but lateral effects are neglected (see [3, 5, 38–43]). Many propagation virtual tools have been introduced over the last decade. Analytical models for relatively simple scenarios are mostly based on mode and ray summation [9, 15, 16]. Powerful time-domain models are based on both the FDTD [13] and the TLM [14] methods. The frequency-domain models include simulators based on the SSPE [5, 10, 34], the FEMPE [44], and the MoM [2, 12, 45]. There are also tools for comparisons of the SSPE and the MoM [31], or of the SSPE and the FEMPE [46].

One of the important issues in these 2D propagation simulation studies is the process of VV&C [8, 45]. The VV&C procedure necessitate reliable reference data. Reliable data are best obtained via measurements. Unfortunately, this is extremely expensive, complex, and time consuming. Hence, one has to generate reference data from analytical models (only after accurate numerical computations), which are available for only a few idealized, simplified canonical propagation problems.

Radio Wave Propagation and Parabolic Equation Modeling, First Edition. By Gökhan Apaydin, Levent Sevgi
© 2017 by the Institute of Electrical and Electronic Engineers, Inc. Published 2017 by John Wiley & Sons, Inc.

Analytical models are mostly based on mode and ray summations and require eigenmode and/or eigenray extraction. Modes are global wave objects in guiding environments and can be used in representing any kind of excitation. They are confined transversely and propagate longitudinally. Modes and their propagation constants are also called eigenfunctions and eigenvalues, respectively. On the other hand, rays are local wave objects and are source–observer dependent. Numerical computation of mode and/or ray summation for different excitations might be a challenge.

Another challenge is accurate source modeling. Different models use different excitations such as a line source, a vertical short dipole, or a directional transmitter (e.g., a vertical Gaussian beam). Vertical (upward/downward) tilt may also be given in the last source model.

Accurate source modeling and VV&C of different 2D propagators are discussed in this chapter. The SSPE, FEMPE [45], and MoM propagators are validated and calibrated against the 2Ray model over flat Earth and the single knife edge problem with the 4Ray model [8].

4.2   Accurate Source Modeling

In most of the 2D propagation models, the source is modeled as a transmitting antenna's pattern. This is achieved in PE propagators by using a vertical Gaussian-shaped input field profile (see Fig. 4.1). The vertical beamwidth and beam tilt can also be adjusted. On the other hand, most of the analytical solutions (i.e., mode/ray models) that may serve as a reference use line-source excitation. In order to compare these two and perform VV&C, the excitations must be identical. Hence, source adaptation is crucial.

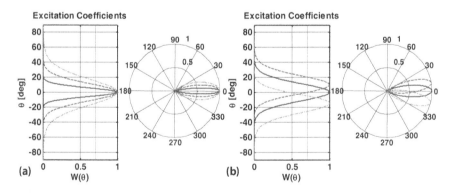

Figure 4.1   Ray excitation coefficients and polar antenna patterns of vertical Gaussian beams with different tilts and their wavenumber equivalents: (a) untilted case: (solid) $\theta_{bw} = 10°$, (dashed) $\theta_{bw} = 20°$, (dashed-dotted) $\theta_{bw} = 30°$; (b) tilted case with $\theta_{bw} = 20°$ (solid) $\theta_{elv} = 0°$, (dashed) $\theta_{elv} = 10°$, (dashed-dotted) $\theta_{elv} = -15°$.

As observed in Fig. 4.1, using a Gaussian function for the vertical source profile gives a smooth beam without any side lobes and back lobes. In practice, vertical

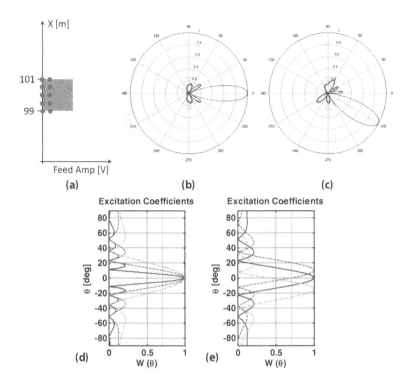

Figure 4.2 (a) A vertical (2×5) planar array, (b) untilted, (c) $30°$ down-tilted beam (the angular distribution is characterized by a *Sinc* function), and the ray excitation coefficients (d) without tilt: (solid) $\theta_{bw} = 10°$, (dashed) $\theta_{bw} = 20°$, (dashed-dotted) $\theta_{bw} = 30°$; (e) $\theta_{bw} = 20°$ with tilt: (solid) $\theta_{elv} = 0°$, (dashed) $\theta_{elv} = 10°$, (dashed-dotted), $\theta_{elv} = -15°$.

arrays, either linear or planar, are used in communication and/or radar systems. A more realistic vertical source model is shown in Fig. 4.2. Here, a 2×5 planar array is shown. The inter-array element distances are 0.5 m and 0.25 m, in the vertical and horizontal directions, respectively. At a frequency of 300 MHz, these distances correspond to half and quarter wavelengths, respectively. The source is 100 m above the ground. The vertical extent of this source is 2 m. The beamwidth is around $30°$ (using $\theta_{bw} \approx \lambda/D$ radian with $\lambda = 1$ m and $D = 2$ m, the beamwidth is found to be $28.6°$). Note that both untilted and tilted beams have side lobes and back lobes, and the side lobes have to be taken into account in realistic excitations. The ray excitation coefficients corresponding to the planar array are shown in Fig. 4.2.

The beamwidth is controlled by the number of transverse elements in the array. The tilt is controlled by element phasing (i.e., electronic beamforming and beam steering). Note that the smooth Gaussian beams presented in Fig. 4.1 can also be produced with the vertical planar array pictured in Fig. 4.2. This could be achieved by controlling/changing element feed voltages. For example, using a binomial feed voltage distribution in Fig. 4.2 will produce exactly the same Gaussian beams given in Fig. 4.1 (instead of using 1 V, 1 V, 1 V, 1 V, 1 V for the five elements in Fig. 4.2, a binomial excitation uses 1 V, 4 V, 6 V, 4 V, 1 V).

The numerical propagators (SSPE, FEMPE) directly use either a Gaussian beam or an actual planar array distribution as an initial field (antenna) profile. If the environmental refractivity variations and/or physical transverse boundaries allow EM wave trapping/guiding, one can use a mode summation approach as an analytical model. Accurate source modeling in this case only necessitates the computation of modal excitation coefficients from orthonormality relations [2, 45]. Unfortunately, this is not the case for the simple flat Earth scenario which is not a guided wave problem; mode summation is therefore not possible.

The simplest solution approach for the flat Earth problem is to apply a ray summation model. As described in Chapter 2, the total field at the observation point is obtained via the addition of the contributions of the direct and ground-reflected waves. This may also be achieved by the image method. An image source is located beneath the ground and the ground is removed. The addition of the contributions of both the source and its image yields the solution. The polarity of the image source is chosen depending on the polarization of the source. The MoM solution is also a ray-based model; accurate source modeling for the 2Ray model is therefore directly applicable to the MoM.

Accurate source modeling in ray-based models under a tilted Gaussian beam (see Fig. 4.1) and/or vertical planar array (see Fig. 4.2) is as follows:

- Choose the vertical field distribution $u(z_s, x)$ for the PE models (for both SSPE and FEMPE). In general, this is a vertically tilted beam and $z_s = 0$.

- Make the source unit power by using the normalization constant obtained from

$$c_x = \left[\int_{x=0}^{\infty} u^2(z_s, x) dx \right]^{-1/2}. \tag{4.1}$$

- Apply the FFT between the vertical spatial, x, and vertical wavenumber, $k_x = k_0 \sin \theta$, domains for free space. This gives the angular distribution of the excitation, $U(z_s, k_x)$ or $U(z_s, \theta)$.

- Add the ray excitation coefficients $W(\theta)$ to the ray with respect to the departure angles. Note that $W(\theta)$ must also be unit power: that is, c_θ must satisfy

$$\int_{\theta=-\frac{\pi}{2}}^{\frac{\pi}{2}} c_\theta^2 W^2(\theta) d\theta = 1. \tag{4.2}$$

The source model can be operated between the x and k_x-domains (as mentioned above) which are the Fourier transform pairs. The domains are determined with respect to their maximum heights and wavenumbers, which are between $[0, X_{max}]$ and $[-k_{xmax}, k_{xmax}]$. The discretization with respect to mesh spacing is determined to be $dx \times dk_x = 2\pi/N$ where N is the transform size. The antenna pattern, $U(z_s, k_x)$, is specified in the transverse wavenumber domain in (3.18), where $f(k_x)$ is either a Gaussian (see Fig. 4.1)

$$f(k_x) = \exp\left(-\frac{k_x^2 \ln 2}{2k_0^2 \sin^2(\theta_{bw}/2)}\right) \tag{4.3}$$

or planar (see Fig. 4.2)

$$f(k_x) = \operatorname{sinc}\left(\frac{k_x D}{2}\right) \tag{4.4}$$

where $D = 2\operatorname{sinc}^{-1}(1/\sqrt{2})/(k_0 \sin(\theta_{bw}/2))$ is the planar array width. The tilt angle (θ_{elv}) is introduced by replacing $f(k_x)$ with $f(k_x - k_0 \sin \theta_{elv})$. The initial field profile, $u(z_s, x)$, is then obtained via the inverse FFT of $U(z_s, k_x)$. The corresponding ray excitation coefficients are then approximated as

$$W(\theta) \approx \exp\left(-\frac{[\sin\theta - \sin\theta_{elv}]^2 \ln 2}{2\sin^2(\theta_{bw}/2)}\right) \tag{4.5}$$

$$W(\theta) \approx \operatorname{sinc}\left(\pi [\sin\theta - \sin\theta_{elv}] \frac{D}{\lambda}\right) \tag{4.6}$$

for every ray, using their departure angles, θ, and these coefficients are used in the ray summation model for the antenna pattern. Note that the excitation coefficients are multiplied with the normalization constant, c_θ, which satisfies (4.2).

4.3 Wave Propagators in Two Dimensions

Wave propagation has long been modeled in two dimensions. Here, the problem is handled in transverse–longitudinal (i.e., height–range) coordinates. The source is located at the initial range (usually, $z_s = 0$) at a specified height ($x_s = h_t$). The receiver is located at the final range ($z_o = d$) at a specified height ($x_o = h_r$). Figure 4.3 shows the simplest 2D propagation environment; flat Earth with PEC ground. The environment is open longitudinally ($z \to \pm\infty$) and semi-open transversely ($x \to \infty$). The top and bottom plots show the same scenario with the line source and directive antenna (tilted-beam) excitations, respectively.

4.3.1 Flat Earth and Two-Ray Model

The simplest propagation scenario used in analytical modeling is the flat Earth with PEC surface under a line source as excitation (see the scenario in Fig. 4.3).

A line source (aligned with the y-axis) allows the reduction of the 3D wave equation into two dimensions. In this case, cylindrical-wave spread is of interest. As pictured in the figure, the total field at the observer is obtained via vector addition of direct and ground-reflected rays. This model is called 2Ray model. Rays in two

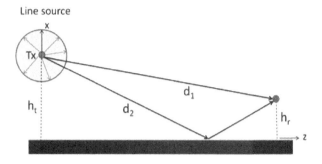

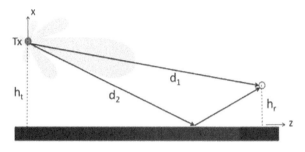

Figure 4.3 A flat Earth with a PEC boundary.

dimensions are attenuated by the square root of the distance as they propagate away from the source. The total field is therefore

$$u(z_o, x_o) \approx W(\theta_1)\frac{\exp(ik_0 d_1)}{\sqrt{d_1}} + W(\theta_2)\Gamma_{h,v}\frac{\exp(ik_0 d_2)}{\sqrt{d_2}} \qquad (4.7)$$

where $W(\theta_1)$ and $W(\theta_2)$ are the ray excitation coefficients to the direct and ground-reflected rays, $(z_o = d, x_o = h_r)$ are the observer coordinates, and

$$d_1 = \sqrt{d^2 + (h_t - h_r)^2}, \; d_2 = \sqrt{d^2 + (h_t + h_r)^2} \qquad (4.8)$$

where d is the longitudinal distance between the source and the observer, and the complex reflection coefficient, $\Gamma_{h,v}$, for the horizontal and vertical polarizations is given in (2.8). The departure angles of the two rays are

$$\theta_1 = \tan^{-1}\left[(h_r - h_t)/d\right], \; \theta_2 = -\tan^{-1}\left[(h_r + h_t)/d\right]. \qquad (4.9)$$

Assuming infinite conductivity (i.e., a PEC surface), $\Gamma_h \to -1$ and $\Gamma_v \to 1$.

Note that the ray excitation coefficients are necessary if a directive antenna with a given vertical beamwidth and vertical tilt is used as pictured at the bottom of Fig. 4.3. A MATLAB code *Fearth1.m* uses 2Ray model with ray excitation coefficients.

4.3.2 FEM-Based PE Wave Propagator

A finite-element PE is a one-way, forward propagation model, valid under the paraxial ($+z$ direction) approximation. The FEMPE is based on the division of the vertical profile between lower and upper boundaries into nodes. A vertical antenna profile, $u(z_s, x)$, is located at the selected discrete nodes and propagated longitudinally using the Crank–Nicolson method, based on improved Euler approach [44]. A widely used antenna pattern is the normalized, tilted Gaussian pattern. See Section 3.6 for details. A MATLAB code *Fearth2.m* uses FEMPE model with ray excitation coefficients.

4.3.3 SSPE-Based PE Wave Propagator

The SSPE model has long been used in groundwave propagation modeling. It models EM wave scattering above irregular, lossy Earth through a non-homogeneous atmosphere. It takes into account only one-way (forward) scattering effects in the paraxial region (i.e., for near-axial angles). The FFT-based PE solution uses a longitudinal marching procedure. First, an antenna pattern representing the initial height profile is injected. The initial field is then propagated longitudinally from z_s to $z_s + \Delta z$, and the transverse field profile at the next range is obtained. This new height profile is then used as the initial profile for the next step, and the procedure goes on until the propagator reaches the desired range. See Section 3.5 for details. A MATLAB code *Fearth3.m* uses SSPE model with ray excitation coefficients.

4.3.4 Method of Moments Modeling

The MoM technique can be used to find the propagation of horizontally and vertically polarized waves by using the electric field integral equation and the magnetic field integral equation, respectively. Open-region propagation over the irregular ground and/or rough surfaces has been successfully modeled with the MoM. In the classical MoM, the integral equation is converted to the corresponding matrix equation via the discretization of the ground/surface. An $N \times N$ system of equations, $[V] = [Z][I]$, is then constructed, and numerically solved. Here, $[I]$ contains the unknown segment currents, $[V]$ contains segment voltages excited by the source, and $[Z]$ is the $N \times N$ impedance matrix of the ground/surface. A solution of this system yields the unknown segment currents. Superposition of the contributions of the segment currents via Green's function of the problem yields the ground scattered field. Finally, the total field is obtained by adding the incident field [47]. A MATLAB code *Fearth4.m* uses MoM model with ray excitation coefficients, that is compared with the 2Ray model for accurate source modeling.

4.4 Knife Edge and Four Ray Model

The single knife edge problem is a canonical structure where reference data can be generated for VV&C procedures. The ray summation approach is based on the construction of four different rays, and related reflection and diffraction coefficients; Fresnel integrals are also used (See Section 2.2 for details).

The total field for 4Ray model with ray excitation coefficients is obtained as

$$F = \sum_{q=1}^{4} W(\theta_q)\Gamma_q \frac{1-i}{2} \exp\left(ikR_q\right) \{(0.5 + C(v_q)) + i\,(0.5 + S(v_q))\} \quad (4.10)$$

where $W(\theta_q)$ is the ray excitation coefficient to be determined from the antenna pattern; other parameters are defined in Section 2.2. The departure angles of the four rays are $\theta_1 = \theta_4 = \tan^{-1}\left[(h_r - h_t)/d\right]$ and $\theta_2 = \theta_3 = -\tan^{-1}\left[(h_r + h_t)/d\right]$.

A MATLAB code *Knife_SSPE4Ray.m* uses the 4Ray model with ray excitation coefficients for source modeling.

4.5   Canonical Tests and Calibration

First, fields vs. range/height variations are mapped in color plots. Figure 4.4 shows 3D color maps for untilted and tilted beams. The frequency is 30 MHz and the vertical beamwidth is $30°$. The source is located 200 m above the ground. The longitudinal and transverse mesh sizes are $\Delta z = 1$ m and $\Delta x = 5$ m, respectively.

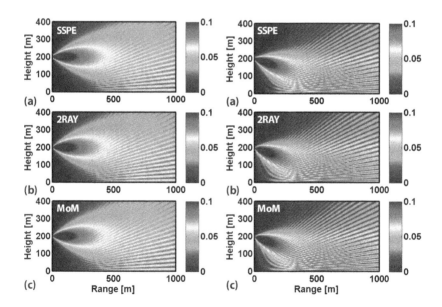

Figure 4.4   Fields vs. range/height for a given (left) untilted (right) $-15°$-tilted Gaussian source: (a) SSPE, (b) 2Ray, (c) MoM (one-way propagation, DBC, $f = 30$ MHz, source height is 200 m, $\theta_{bw} = 30°$, $\Delta z = 1$ m, $\Delta x = 5$ m).

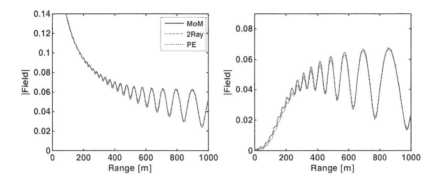

Figure 4.5 Fields vs. range for a given untilted Gaussian source (at 200 m height, observer height is (left) 200 m and (right) 100 m, $\theta_{bw} = 30°$, $f = 30$ MHz, $\Delta z = 1$ m, $\Delta x = 5$ m).

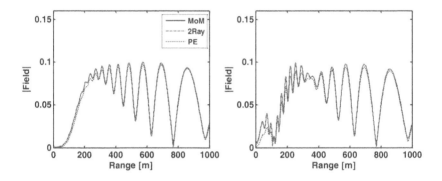

Figure 4.6 Fields vs. range for a given tilted (left) Gaussian source and (right) planar array (at 200 m height, observer height is 100 m, $\theta_{bw} = 20°$, $\theta_{elv} = -15°$, $f = 30$ MHz, $\Delta z = 1$ m, $\Delta x = 5$ m).

In order to observe the degree of the matching of the results of the models, 2D variations in fields vs. range are plotted in Fig. 4.5. As observed, very good agreement is obtained among the models without any normalization at the observer point.

Figure 4.6 shows fields vs. range variations for both a tilted Gaussian source and a planar array. The effects of the excitations in these plots are clearly observed.

The flat Earth scenario obviously is the simplest propagation scenario for the analytical ray model. Unfortunately, even this simplest scenario may be challenging in terms of numerical modeling.

- Both SSPE and FEMPE are restricted with the paraxial propagation. Their narrow- and wide-angle versions can handle waves propagating with vertical angles up to $\pm15°$ and $\pm45°$, respectively. This means that, depending on the

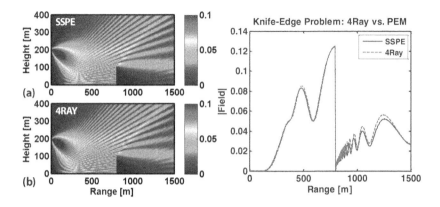

Figure 4.7   (Left) Fields vs. range/height, (right) fields vs. range at 80 m height for a given tilted Gaussian source using (a) SSPE, (b) 4Ray model (one-way propagation, DBC, $f = 30$ MHz, with 100 m high knife edge at 800 m range, source height is 200 m, $\theta_{bw} = 30°$, $\theta_{elv} = -15°$, $\Delta z = 1$ m, $\Delta x = 5$ m).

source height, they are invalid at short ranges. The restrictions may become more serious if vertically tilted beams are used ($|\theta_{elv} + \theta_{bw}/2| \leq 45°$).

- The MoM suffers abrupt longitudinal truncation of the segments at both ends. Although the propagation medium extends to infinity horizontally, regions (i.e., segments) before the first segment and after the last segment are neglected. Because of this, the MoM results near the source and near the observer will be erroneous. If results near the source and observer are important, then one needs to use a number of additional segments on both ends (a few wavelength extension would be enough).

Nevertheless, simulations are repeated for the knife edge problem with similar scenarios. The second scenario belongs to a PEC surface and a knife edge illuminated by a Gaussian source through the homogeneous atmosphere at 30 MHz. Figure 4.7 plots 3D field maps generated via SSPE and 4Ray packages. The scenario belongs to one-way propagation for horizontal polarization. The heights of the knife edge obstacle at 800 m range are 100 m. The Gaussian source is at 200 m height. The excellent visual agreement shows the power of the accurate source modeling.

CHAPTER 5

PE AND TERRAIN MODELING

5.1 Irregular PEC Terrain

Irregular terrain modeling can be implemented in the PE via several different mathematical approaches and it is possible for the user to choose the appropriate one for
his/her problem [2, 31, 48, 49].

The *staircase approximation* of the range dependent terrain profile is the best and
easiest in order to handle DBC since neither analytical terrain function nor slope
values are required; only the terrain height at each range step is needed. In this approach, on each segment of the constant height, the vertical field profile is calculated
in the usual way, applying the desired BCs on the ground surface, and then, the field
values are simply set to zero inside the terrain. This approach provides satisfactory
results in the approximate sense because the BCs on sloping facets are not properly
taken into account, and also, the corner diffraction is ignored. Figure 5.1 shows
the staircase approximation of irregular terrain. By choosing small Δz and Δx, the
model approaches to the exact irregular terrain.

There are more accurate approaches in the literature to the modeling of sloping
facets by representing the terrain as a sequence of piecewise linear functions, or
by employing coordinate transformations, which achieve conformal mappings over

Radio Wave Propagation and Parabolic Equation Modeling, First Edition. By Gökhan Apaydin, Levent Sevgi
© 2017 by the Institute of Electrical and Electronic Engineers, Inc. Published 2017 by John Wiley & Sons, Inc.

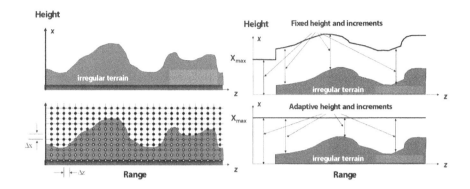

Figure 5.1   (Left) Field calculation of irregular terrain using (right-top) coordinate transformation, (right-bottom) staircase approximation.

terrain curvature. Neumann and Cauchy BCs necessitate the use of *coordinate transformation* which is nothing but a simple coordinate transform $z' \to z$, $x' \to x - t(z)$ where $t(z)$ is the longitudinal terrain function along the propagation direction. Coordinate transformation is used to handle irregular terrain effects. Irregular terrain can be introduced either via a terrain height function $t(z)$ or a terrain file which contains range–height data pairs.

For the implementation of PE, the second-order derivative of terrain function, $t''(z)$, is included in the refractive index. Note that only forward scattering effects are taken into account in PE with this implementation. Simply, the second normal derivative of the terrain function is added to the refractive index term to obtain irregular terrain effects.

When the coordinate transformation is used, the calculated fields are always above ground, therefore the fields above maximum height should be eliminated. If the staircase approximation is used, the calculated fields are above $x = 0$ (see Fig. 5.1 for details).

A MATLAB code *FEMPE.m* uses FEM over irregular terrain. Both the staircase approximation and coordinate transformation are taken into account. Here, the fields inside terrain are set to zero for staircase approximation. On the other hand, the second-order derivative of terrain function (*ddfxter* in *TerFEMPE* function) is added to the refractivity function and $x' \to x - t(z)$ is considered for coordinate transformation.

5.2   PE and Impedance Boundary Modeling

Surface-wave propagation along multi-mixed propagation paths including irregular terrain profiles has long been a challenging wave propagation prediction problem. Early analytical models, such as the Norton (ray-optical) [50] and Wait (normal mode) [51] formulations are valid for homogeneous propagation paths over a smooth spherical Earth with an impedance (Cauchy-type) BC. Propagation effects over multi-mixed smooth paths (e.g., sea–land transitions without taking land irregularities into account) can be predicted via the Millington approach [52]. These have been revisited, well investigated, and well documented because of novel digital communication, radar, and radio systems [3, 4, 9, 15, 16, 53]. Unfortunately, these analytical and numerical models are not capable of handling irregular mixed path effects. There have been a few attempts for non-smooth mixed paths such as bluffs [54], Gaussian-shaped hilly islands [55], but realistic models and systematic predictions have still been missing. Previous studies discussed either mixed (but smooth) path effects or irregular (but homogeneous) terrain modeling. However, none of them included both. (Note that *irregular terrain* refers to the change of land's height above Earth's surface along the propagation direction, with a homogeneous impedance boundary, like the land or sea. On the other hand, *inhomogeneous terrain* means the impedance BC changes with the range, as in a land–sea transition.) For example, Marcus investigated the effects of irregular terrain and an inhomogeneous atmosphere on groundwave propagation in [56]. All of his examples were on a homogeneous but irregular surface. Janaswamy modeled only irregular terrain effects without mixed paths in [57], and irregular and inhomogeneous terrain in [58]. However, he only presented examples for irregular terrain (e.g., a lossy Gaussian hill). There was no inhomogeneous (mixed path) modeling. In [59], Hviid and his friends modeled only irregular terrain effects. Although Levy's book includes almost all groundwave propagation problems, there are only two examples for mixed paths (see Figs. 9.5 and 9.7 on pages 159 and 161 in [5]), mixed path modeling with irregular islands was missing. These are all because surface-wave modeling at LF/MF/HF bands (i.e., from a few 100 kHz up to 30 MHz) along multi-mixed paths with irregular land profiles is extremely challenging, especially in terms of numerical difficulties [34].

A highly attractive and effective model for the prediction of EM wave propagation effects above Earth's surface is the SSPE method. The PE models a one-way, forward propagation problem. It can be numerically discretized and solved via step-by-step iterative marching representations from source to receiver, using the finite-difference, FEM, or MoM approaches as well as DFT algorithms. The SSPE has been widely used for wave propagation modeling above an irregular Earth's surface, through an inhomogeneous atmosphere. It has been used at upper VHF and above (i.e., 100–150 MHz and above), especially at microwave frequencies, where DBC and/or NBC are appropriate [5,18,60,61]. On the other hand, surface-wave propagation modeling along multi-mixed paths at LF/MF/HF necessitates the use of an impedance BC. The DMFT has been added to the SSPE for this purpose [62]. The SSPE has then been extended to general irregular terrain problems [32], accelerated [63], and instabilities removed [33]. The PE method has also been used to represent two-way propagation problems [64–68].

The most challenging problem that still needs to be further investigated in two dimensions is the prediction of EM wave propagation and PL variations for surface-waves above the Earth's surface along multi-mixed (i.e., inhomogeneous) and non-smooth (i.e., having an irregular terrain profile) paths. This has been addressed lately and powerful numerical simulators have been introduced [31,44,48]. First, a systematic investigation with characteristic applications and reliable comparisons against SSPE was given in [31]. There, a numerical propagator based on the FEM was developed, tested and calibrated against analytical exact (reference) data as well as against available SSPE propagators. FEM-based surface-wave propagation package was then introduced [44], which accounts for the signal attenuation along multi-mixed propagation paths through a homogeneous (standard) atmosphere. This package was also tested against International Telecommunication Union (ITU) curves as well as against the Millington curve fitting method [52]. Finally, surface-wave attenuation in the LF/MF/HF bands, due to successively isolated islands aligned along sea/ocean propagation paths, was discussed in [48].

The *DMFT-based SSPE algorithm*, which accounts for mixed paths with impedance BCs, is combined with the coordinate-transformation-based irregular terrain algorithm in [34]. Sea–land–sea paths with irregular land pieces are numerically modeled. A MATLAB-based numerical propagator is developed for this purpose. The surface-wave attenuation due to hilly islands along multi-mixed propagation paths is precisely predicted. Tests are done, and comparisons against the Millington package [15] and the FEMPE algorithm with mixed paths (FEMIX) [44, 48] are given (see Chapter 10 for Millington and FEMIX packages).

5.2.1   Discrete Mixed Fourier Transform (DMFT)

The DMFT approach is applied to the discretized vertical field $u(z, m\Delta x)$ for $m = 0, ..., N$. A patch is added to the SSPE algorithm to satisfy the impedance BC at the surface [59]

$$U(z,0) = A \sum_{m=0}^{N} r^m u(z, m\Delta x) \tag{5.1}$$

$$U(z, l\Delta k_x) = A \sum_{m=0}^{N} \left[\begin{array}{c} \alpha_2(z) \sin\left(\dfrac{\pi l m}{N}\right) \\ -\dfrac{1}{\Delta x} \sin\left(\dfrac{\pi l}{N}\right) \cos\left(\dfrac{\pi l m}{N}\right) \end{array} \right] u(z, m\Delta x) \tag{5.2}$$

$$U(z, N\Delta k_x) = A \sum_{m=0}^{N} (-r)^{N-m} u(z, m\Delta x) \tag{5.3}$$

for $l = 1, ..., N - 1$, where $A = 2(1 - r^2)/\left((1 + r^2)(1 - r^{2N})\right)$; Δk_x is the step size to satisfy $\Delta k_x \Delta x = \pi/N$; r and $-1/r$ are the roots of the quadratic equation, $r^2 + 2r\alpha_2(z)\Delta x - 1 = 0$, for vertical and horizontal polarizations, respectively. The summations in (5.1)–(5.3) use weighted factor of 0.5 for the initial and last terms ($m = 0$, $m = N$). The DMFT of the field at the next range step ($z + \Delta z$) is then obtained using

$$U(z + \Delta z, 0) = \exp\left(\frac{i\Delta z}{2k_0} \left(\frac{\log r}{\Delta x}\right)^2\right) U(z, 0) \tag{5.4}$$

$$U(z + \Delta z, l\Delta k_x) = \exp\left(i\Delta z\left(\sqrt{k_0^2 - (l\Delta k_x)^2} - k_0\right)\right)U(z, l\Delta k_x) \quad (5.5)$$

$$U(z + \Delta z, N\Delta k_x) = \exp\left(\frac{i\Delta z}{2k_0}\left(\frac{\log(-r)}{\Delta x}\right)^2\right)U(z, N\Delta k_x) \quad (5.6)$$

for $l = 1, ..., N - 1$. Finally, the field $u(z + \Delta z, m\Delta x)$ for $m = 0, ..., N$ at the next range step is calculated by using the inverse DMFT

$$u(z + \Delta z, m\Delta x) = U(z + \Delta z, 0)r^m + U(z + \Delta z, N\Delta k_x)(-r)^{N-m}$$
$$+\frac{2}{N}\sum_{i=0}^{N}\left(U(z + \Delta z, l\Delta k_x)\frac{\alpha_2(z)\sin\left(\frac{\pi lm}{N}\right) - \frac{1}{\Delta x}\sin\left(\frac{\pi l}{N}\right)\cos\left(\frac{\pi lm}{N}\right)}{\alpha_2^2(z) + \frac{1}{\Delta x^2}\sin^2\left(\frac{\pi l}{N}\right)}\right) \quad (5.7)$$

for $m = 0, ..., N$. Note that it is possible to eliminate (5.2) and simplify the method by using the DST of a newly defined function in terms of $u(z, m\Delta x)$ and $\alpha_2(z)$ [5].

The PE represents an initial-value problem, and a Gaussian antenna pattern with the initial field's height profile $u(z_s, x)$ is injected.

The flowchart of the DMFT-SSPE algorithm is given in Fig. 5.2. The implementation of the DMFT-SSPE algorithm with the coordinate transformation is as follows:

- The initial field profile $u(z_s, m\Delta x)$ is injected.

- The coordinate transformation is applied to implement irregular terrain profiles within the SSPE algorithm, using a terrain file.

- The second derivative of the terrain function is added to the refractive index term.

- The DMFT is applied via (5.1)–(5.3) using $\alpha_2(z)$ and $U(z)$ is obtained.

- At the next range step, $U(z + \Delta z)$ is obtained from $U(z)$ via (5.4)–(5.6).

- Finally, the field at the next range, $u(z + \Delta z)$, is obtained via the application of the inverse DMFT using (5.7).

MATLAB codes, *DMFTSSPE.m* and *FEMPE.m*, use DMFT-SSPE and FEMPE over irregular terrain, respectively. Here, function *deterslp* determines the first- and second-order derivatives of terrain functions of Gaussian- and irregular terrain-shaped islands. The DMFT algorithm is available in *SSPE* function and the input specification is in *PEINP.DAT* file.

5.3 Numerical Results and Comparison

Both SSPE and FEMPE can handle propagation problems above the PEC or lossy irregular terrain profiles through the variable atmosphere.

First scenario presented in Figs. 5.3 and 5.4 belong to the tests and comparisons over different user specified irregular PEC terrain profiles through the atmosphere

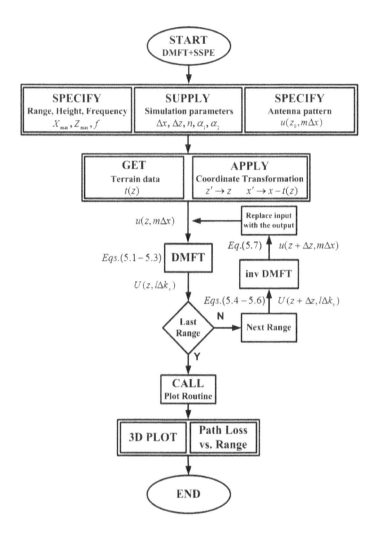

Figure 5.2 The flowchart of the DMFT-SSPE algorithm.

with piecewise bilinear and trilinear vertical refractivity variations under horizontal polarization. The maximum range on the left side of Fig. 5.3 is 60 km, the frequency is 100 MHz, the source height is 700 m, the tilt angle is 2° downward, and the beamwidth is 1°. The refractivity is bilinear, the refractivity slope is 2100 M units/km and linearly increasing up to 700 m, and then decreasing with the same slope. The elevated duct is clearly observed. Secondly, the maximum range on the right side of Fig. 5.3 is 50 km, the frequency is 100 MHz, the source height is 250 m, the tilt angle is 2° upward, and the beamwidth is 1°. The refractivity at this time is trilinear, the refractivity slope is 5000 M units/km and linearly decreasing up to 400 m,

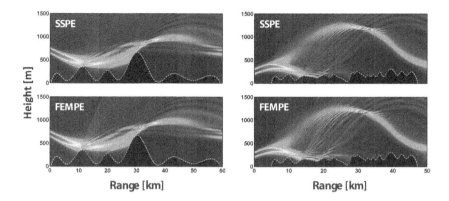

Figure 5.3 3D fields vs. range/height variations through (left) bilinear, (right) trilinear atmosphere over an irregular PEC terrain (top) SSPE, (bottom) FEMPE results (staircase approximation, horizontal polarization).

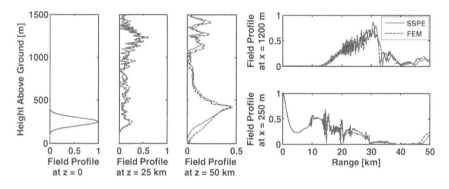

Figure 5.4 (Left) Vertical fields vs. height at three specified ranges, (right) horizontal fields vs. range at two specified heights through a trilinear atmosphere over an irregular PEC terrain; (solid) SSPE, (dashed) FEMPE.

linearly increasing between 400 m and 800 m, and then linearly decreasing with the same slope. Figure 5.4 shows 2D vertical and horizontal field profiles through a trilinear atmosphere for comparison.

For the comparison of polarization, the SSPE propagator is used to investigate various complex propagation problems. Various PEC irregular terrain paths are generated and propagation above irregular terrain through the standard atmosphere (including the Earth's curvature) is simulated under both horizontal polarization and vertical polarization. 3D fields vs. range/height plot at 300 MHz through a standard atmosphere is pictured in Figs. 5.5 and 5.6. Only, the SSPE map is shown but the FEMPE map is also the same. The maximum range is 100 km, the source height is 400 m, the tilt angle is 0.5° downward, and the beamwidth is 0.5°. As observed,

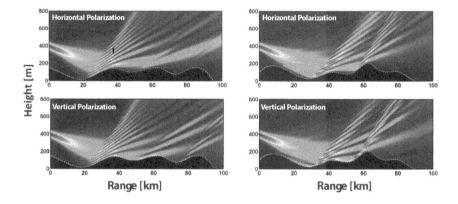

Figure 5.5 Three-dimensional fields vs. range/height variations produced with SSPE at 300 MHz through a standard atmosphere over different irregular PEC terrains: (top) horizontal polarization, (bottom) vertical polarization (coordinate transformation).

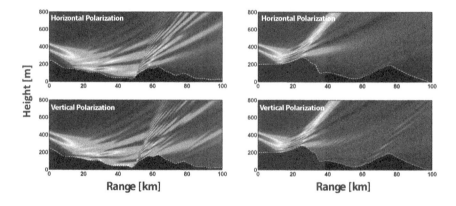

Figure 5.6 Three-dimensional fields vs. range/height variations produced with SSPE at 300 MHz through a standard atmosphere over different irregular PEC terrains: (top) horizontal polarization, (bottom) vertical polarization (coordinate transformation).

down the propagation of the beam, reflection from the terrain and interference between the direct and terrain-reflected waves are clearly observed.

Related to the lossy irregular terrain, Fig. 5.7 belongs to a 40 km homogeneous path with sea surface parameters $\sigma = 5$ S/m, $\varepsilon_r = 80$ having a 10 km long, 250 m high hills. Fields vs. range/height plot at 10 MHz are produced with FEMPE. Here, the surface-wave energy accumulates in the front slope of the hill and the signal strength increases. The example using a full sinusoidal hill represents a tsunami wave on the sea/ocean with these electrical parameters, or a mountain on a propagation path over land if electrical parameters are changed to land values. Figure 5.7 shows how energy accumulates in the front slope of the hill. It requires concave/convex

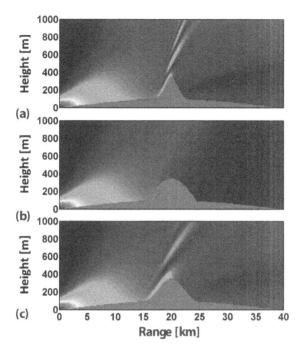

Figure 5.7   Three-dimensional fields vs. range/height produced with FEMPE at 10 MHz through a standard atmosphere over different irregular lossy terrains with vertical polarization; (a) Gaussian-shaped hill, (b) half sinusoidal hill, (c) full sinusoidal hill (the hills are 10 km long and 250 m high with surface parameter $\sigma = 5$ S/m and $\varepsilon_r = 80$).

irregularity in general to accumulate energy. Observe that half sinusoidal hill is fully convex and the full sinusoidal hill is partially concave/convex at both sides so the surface-wave energy accumulates in the front slope of the hill.

The next scenario belongs to triple tests performed among the SSPE package, the Millington package [15] (prepared according to ITU Recommendations), and FEMIX package [44, 48] (see Chapter 10 for details). PL vs. range along a three-segment 40 km long mixed path, predicted via the three packages, is given in Fig. 5.8. In this scenario, there is a 10 km long flat island 15 km away from the transmitter on the left side. On the top, 3D fields vs. range/height at 5 MHz, produced with the SSPE package, is given. At the bottom, PL vs. range at different frequencies is plotted. The surface parameters are $\sigma = 0.002$ S/m, $\varepsilon_r = 10$ for island and $\sigma = 5$ S/m, $\varepsilon_r = 80$ for sea. As observed, the surface-wave detaches from the surface as it propagates because of the Earth's curvature. A small portion follows the surface as the surface-wave. The sea–land discontinuity causes an extra sharp detachment (energy tilt up). This tilt up explains the sharp attenuation first mentioned by Millington. Signal recovery also occurs at the land–sea discontinuity. The right side of Fig. 5.8 belongs to the same scenario with a 250 m high Gaussian-shaped hilly island. Observe that both the SSPE and FEMIX computations account for the

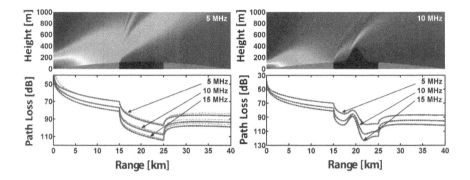

Figure 5.8   (Top) 3D fields vs. range/height produced with SSPE, (bottom) PL vs. range over a three-segment 40 km mixed path, a 10 km long, (left) flat, (right) 250 m high Gaussian-shaped, island is 15 km away from the transmitter with vertical polarization, (solid) FEMIX; (dashed) DMFT-SSPE; (dots) Millington.

attenuation of the surface-wave along the first sea path, the sharp increase in the attenuation at the sea–land discontinuity, signal recovery in the front slope of the island, the additional signal attenuation at the back slope of the island, and finally the signal recovery at the land–sea discontinuity.

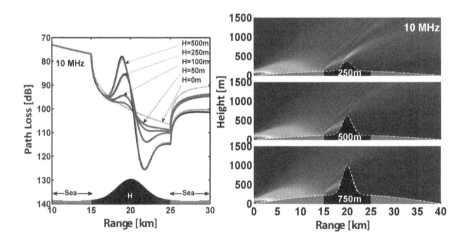

Figure 5.9   (Left) PL vs. range through a standard atmosphere over a three-segment 30 km mixed path (Gaussian-shaped islands with different heights are 15 km away from the transmitter) at 10 MHz with vertical polarization, (solid) FEMIX; (dashed) DMFT-SSPE; (dots) Millington-flat island), (right) 3D signal vs. range/height produced with FEMPE.

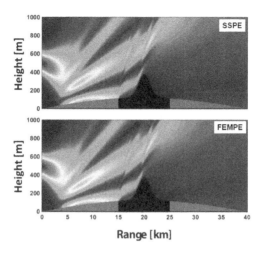

Figure 5.10   Three-dimensional fields vs. range/height produced with SSPE and FEMPE at 10 MHz with vertical polarization through a standard atmosphere along three-segment 40 km mixed path (a 10 km long, 250 m high Gaussian-shaped island is 15 km away from the source).

Figure 5.9 compares the effect of island height over Gaussian-shaped land. PL vs. range over the same scenario at 10 MHz for different island heights is given. As observed, energy accumulation in the front slope and deep signal loss at the back slope of the island hill increase with the hill's height. 3D fields along a three-segment 40 km mixed path are also given, that assists to explain visually the physical phenomenon occurring there. Observe surface-wave propagation, energy detachment because of the Earth's curvature, energy tilt-up at the sea–land discontinuity, and energy accumulation in the front slope of the island.

The next example in Fig. 5.10 belongs to wave propagation over a three-section mixed path (sea–land–sea) with a Gaussian-shaped hilly island at 10 MHz. Here, an elevated antenna is used (a Gaussian antenna pattern with 5° vertical beamwidth, tilted 2° downward, located 500 m above the sea surface) is used to excite surface-waves. As observed, waves hit the sea surface around 5 km; energy couples to the surface and propagates thereafter.

The final example in Fig. 5.11 belongs to an island with an irregular terrain profile. The SSPE is compared against FEMIX for a three-segment 40 km long mixed path at 10 MHz. Excellent agreement between the SSPE and FEMIX results is clearly obtained for PL vs. range.

Although examples presented here are for less than 100 km, ranges for surface-wave propagation in the MF/HF bands extend up to 300–400 km. This could certainly be achievable but would require highly effective absorbing boundary termination in the DMFT-SSPE algorithm with coordinate transformation above the heights of interest.

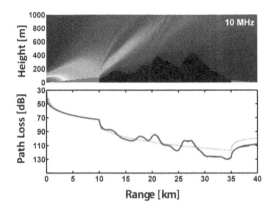

Figure 5.11   (Top) 3D fields vs. range/height produced with SSPE, (bottom) PL vs. range over a three-segment 40 km mixed path through a standard atmosphere over an irregular lossy terrain, a 25 km long arbitrary-shaped island is 10 km away from the transmitter with vertical polarization, (solid) FEMIX; (dashed) DMFT-SSPE; (dots) Millington.

Roughly speaking, the computational time of the DMFT-SSPE algorithm with co-ordinate transformation is at least 8–10 times longer compared to the standard SSPE. Several improvements, such as matrix memory pre-allocation, a vectorization procedure, and efficient memory allocation are used in MATLAB coding to maximize the performance. Some acceleration commands and procedures are also used and the difference in computation times is reduced to less than a factor of three.

The DMFT-SSPE algorithm with the coordinate transformation modeling requires optimization of the discretization parameters according to the frequency and the terrain profile. The maximum height is critical in eliminating reflections from the top boundary and in reducing the computations. Second, the range step should be optimized. Large steps cause phase errors and introduce non-physical oscillations; small steps increase the computation burden. The optimum parameters used here are Δx between 5 m and 20 m, Δz between 5 m and 40 m for various terrain structures and frequencies.

CHAPTER 6

ANALYTICAL EXACT AND APPROXIMATE MODELS

6.1 Wave Propagation in a Parallel Plate Waveguide

Electromagnetic problems (from nanoscale to kilometer-wide systems) are complex in nature but the theory is well established with Maxwell equations (a few of many classical EM books may be listed as [69–71]). The strategies for the solution of such problems may be grouped into three; analytical modeling, numerical simulations, and measurements. Measurements in EM are time consuming, expensive, and, in most cases, extremely difficult to do. A limited number of analytical solutions are available only for a few, highly idealized problems, therefore numerical simulation becomes the only mean for almost all real-life EM engineering problems. This is why EM modeling and simulation (EM-MODSIM) has made a significant progress for the last couple of decades. EM-MODSIM has been taught for sometime worldwide in many universities. A few books about EM-MODSIM may be listed as [3, 72–76]. EM-MODSIM approaches may be divided into two; analytical-model-based and numerical-model-based approaches. The critical issue in EM-MODSIM is the (model) validation, (data) verification, and (code) calibration [2]. The best VV&C method is to compare numerical models with accurately computed analytical models. In this content, high-frequency asymptotics (HFA) plays an essential role in generating reference data [19, 77–80]. HFA models such as GO, geometric theory of diffraction (GTD), physical optics (PO), physical theory of diffraction (PTD), and uniform theory of diffraction (UTD), with their advantages/disadvantages and regions of validity not only yield reference data, but also give physical insight into understanding EM wave scattering pieces. The two HFA tutorials published in [81, 82] review fundamental diffraction models.

In terms of solution approaches, EM problems may be grouped into three: (i) radiation/antenna problems, (ii) scattering and radar cross-section (RCS) modeling, and (iii) waveguiding structures. These all start with Maxwell equations and the definition of BCs. For (i), it is better to define auxiliary scalar and vector potential functions and to solve the equations that contain these auxiliary functions. This is better because components of the vector potential fit in practical excitations. The scattered field representation is used for RCS modeling in (ii) because the excitation (i.e., the incident field) is elsewhere. Finally, Maxwell equations are decomposed into transverse and longitudinal components for guided wave problems in (iii). This not only yields decomposed fields but also scalarizes and solves the problem. Eigenfunctions and eigenvalues are derived from transverse components; excitation coefficients are obtained from longitudinal components. Completeness and orthonormalization relate these two via the well known Sturm–Liouville equation [83]. Most EM-MODSIM studies, commercial or educational, are based on one of the listed models.

- The MoM discretizes the object under investigation into N pieces, called segments or patches, and constructs an $N \times N$ system of equations using Green's function of the problem [84]. This approach is closed form and stable but suffers from a lack of large memory and high-speed computers.

- The PE method, based on either FEM or SSPE discretization, is widely used in complex propagation modeling [5].

- The FDTD method discretizes the physical environment into small cells by replacing partial derivatives with their finite-difference equivalents [85–87]. It is open form and iterative therefore suffers from stability.

- The TLM method uses the circuit equivalent of Maxwell equations [88, 89]. This is also open form and iterative with similar stability problems.

This chapter reviews fundamental EM-MODSIM concepts and the widely used numerical (PE and MoM) models through a reliable VV&C procedure. The philosophy is to keep it as simple as possible so that it can be used as a basic computational electromagnetics note and to supply simple MATLAB codes so that, even beginners can use them. In order to achieve these tasks, one of the most simple and widely used structures is chosen as the EM-MODSIM scenario: propagation inside a parallel plate waveguide with PEC boundaries in two dimensions [90, 91]. The same can also be applied to a wedge waveguide [92].

Propagation inside a parallel plate waveguide is an interesting EM problem where both analytical and numerical models can be tested one against the others.

- First of all, the structure in two dimensions is non-physical; and acts as a low-pass filter.

- Green's function solution (i.e., the EM response of a line source) is exact but requires an infinite number of eigenfunction (mode) summations [90]. This is a numerical challenge especially in the near vicinity of the line source, but it is excellent for the analysis of truncation errors.

- Eigenfunctions (modes) are grouped into two types; propagating modes with real eigenvalues and evanescent modes with imaginary eigenvalues. The number of propagating modes depends on the frequency and the width of the waveguide.

- A tilted directional antenna can also be located inside and can be modeled in terms of modes, but the modal excitation coefficients are now complex. This is another numerical challenge, especially at high frequencies when the number of propagating modes is high.

- The modes (eigenfunctions) are global, therefore they do not suffer from local problems, but extraction of modal excitation coefficients is crucial when generating reference solutions.

- An analytical exact solution can also be constructed in terms of rays which are local wave pieces; again an infinite number of ray summation is required for the line-source excitation. This may be achieved via either eigenray specification or ray shooting. The solution of eigenray equations is difficult especially for rays with a high number of reflections from upper and lower boundaries [90]. Ray shooting is very time consuming and is highly sensitive to source/observer locations. Their computer implementations are both numerical challenges.

- Another ray-based analytical solution is the image method (IM). The IM overcomes the numerical difficulties of both eigenray extraction and ray shooting approaches but a very high number of images are required for accurate field computations.

- One-way, FEM and/or FFT-based PE models have been used in propagation modeling in waveguiding environments [2, 31, 93]. The PE models can handle tilted and directed excitation antennas easily. However, one needs to use wide-angle PE models since narrow-angle models suffer paraxial propagation restriction (i.e., it is valid up to $10°–12°$ vertical propagation angles).

- The two powerful time-domain (FDTD and TLM) methods may also be used in modeling propagation inside a parallel plate waveguide. This is possible even for very long ranges by using sliding window approaches [13, 14]. Both line source and short dipoles can be used in wave excitation in these models, but special care is essential for modeling a tilted and directional transmitting antenna.

▪ Finally, the MoM can be used in propagation modeling inside a parallel plate waveguide. This also requires a special treatment because the MoM has rarely been used in resonating structures (a novel approach, called Mi-MoM, is introduced in [46]). The MoM is based on the solution of an $N \times N$ system of equations that yields currents induced on the segments. The procedure is first applied and currents induced by the source are obtained. Consecutive applications yield segment currents induced by other segment currents. This has to be continued until their contributions diminish.

▪ A hybrid, MoM plus image method (MoM-IM) increases both accuracy and computation time. In this case, an $N \times N$ system of equations is solved once for segment currents induced by the source. The rest is handled via the IM.

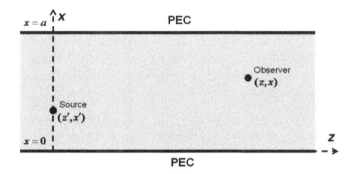

Figure 6.1 A parallel plate waveguide and source–observer locations.

The 2D parallel plate waveguide is pictured in Fig. 6.1. Here, x and z are the transverse and longitudinal coordinates, respectively. The structure is infinite along the y-direction. The height of the waveguide is a. The PEC boundaries are assumed for DBCs for the TE_z (transverse electric with respect to z) problem and for NBCs for the TM_z (transverse magnetic with respect to z) problem (see [1] for TE/TM discussions).

6.2 Green's Function in Terms of Mode Summation

Green's function problem associated with both the TE_z set and the TM_z set is defined by the equation

$$\left\{ \frac{\partial^2}{\partial x^2} + \frac{\partial^2}{\partial z^2} + k_0^2 \right\} g\left(x, z; x', z'\right) = -\delta(x - x')\delta(z - z') \tag{6.1}$$

with BCs

$$g(x, z; x', z') = 0 \ \text{ at } x = 0, a \ (\mathrm{TE}_z) \tag{6.2}$$

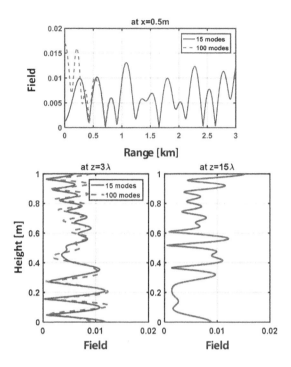

Figure 6.2   (Top) Magnetic fields vs. range, (bottom) magnetic fields vs. height with Green's function: TM_z polarization, $a = 1$ m, $z' = 0$, $x' = 0.3$ m, $k_0 a = 50$, (solid) only propagating modes (15 modes), (dashed) 100 modes.

$$\frac{\partial}{\partial x} g(x, z; x', z') = 0 \ \text{at} \ x = 0, a \ (\text{TM}_z) \tag{6.3}$$

$$g(x, z; x', z') = 0 \ \text{as} \ z \to \pm\infty. \tag{6.4}$$

Here, (x', z') and (x, z) specify the source and observation points, respectively, $\delta(x)$ and $\delta(z)$ are the Dirac delta functions. Green's function $g(x, z; x', z')$ can be obtained as

$$g(x, z; x', z') = \widetilde{g}(z; z') - \frac{2}{a} \sum_{m=1}^{\infty} \frac{\exp(ik_{zm} |z - z'|)}{2ik_{zm}} B(k_{xm} x) B(k_{xm} x') \tag{6.5}$$

$$\widetilde{g}(z; z') = 0, \ B(x) = \sin(x) \ (\text{TE}_z) \tag{6.6}$$

$$\widetilde{g}(z; z') = -\frac{1}{a} \frac{\exp(ik_0 |z - z'|)}{2ik_0}, \ B(x) = \cos(x) \ (\text{TM}_z) \tag{6.7}$$

where $k_{xm} = m\pi/a$, $k_{zm} = \sqrt{k_0^2 - k_{xm}^2}$. The line-source-excited fields are then given by either $E_y = -i\omega\mu_0 g$ or $H_y = -i\omega\varepsilon_0 g$ for TE_z case and TM_z case, respectively.

A short MATLAB code, *PPlate.m*, is prepared for the calculation of the field distribution inside the parallel plate waveguide in terms of mode summation for both polarizations. An example is shown in Fig. 6.2. Here, fields vs. range inside a 1 m wide plate at 0.5 m is pictured. The source is at 0.3 m height. The number of propagating modes for the sets of parameters listed in the figure is 15. The two curves belong to 15 mode and 100 mode summations. As observed, at a distance beyond 0.6 m (i.e., after five wavelengths in distance), only propagating modes contribute. Figure 6.2 also displays fields vs. height at two different distances inside the same plate. As observed, the contribution of only propagating modes at a distance of three wavelengths is not enough to build the correct field.

6.3   Mode Summation for a Tilted Gaussian Source

A directive antenna is used in many propagation applications. This is modeled by using a vertical, tilted Gaussian function in analytical and numerical simulations. This tilted Gaussian source inside a PEC parallel plate waveguide at $z = z'$ may be represented in terms of a modal summation as

$$f(x, z') = \sum_{m=m_0}^{M} c_m(z') v_m B(k_{xm} x) \tag{6.8}$$

where M is the highest mode that should be included for the specified excitation, v_m is the normalization constant

$$v_m = \left(\int_{x=0}^{a} B^2(k_{xm} x) dx \right)^{-1/2} \tag{6.9}$$

and $c_m(z')$ is the modal excitation coefficient, numerically derived from the vertical orthonormality condition as

$$c_m(z') = v_m \int_{0}^{a} f(x, z') B(k_{xm} x) dx. \tag{6.10}$$

The initial field profile $f(x, z')$ at $z' = 0$ is generated from a tilted Gaussian pattern

$$f(x, 0) = \exp\left(ik_0 x \sin \theta_{elv} - \frac{(x - x')^2}{w^2} \right) \tag{6.11}$$

where $w = \sqrt{2 \ln 2}/(k_0 \sin(\theta_{bw}/2))$. The tilted Gaussian antenna pattern is specified by its vertical position (x'), beamwidth (θ_{bw}), and tilt (elevation) angle (θ_{elv}). Note that B again shows either a sine or cosine function behavior, starting from either $m_0 = 1$ or $m_0 = 0$ for horizontal (TE$_z$) and vertical (TM$_z$) polarizations, respectively. The number of modes would be finite for numerical computation. It is common to choose a vertically extending Gaussian function with the arbitrary location, having a vertical elevation angle in the range of $\pm 90°$. Note that the modal

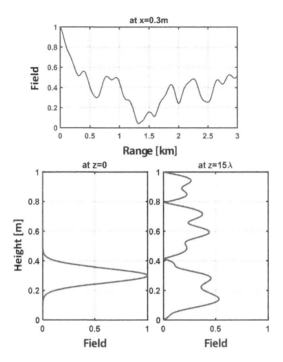

Figure 6.3 (Top) Electric fields vs. range, (bottom) electric fields vs. height with the mode summation ($M = 40$) for a tilted Gaussian source: TE$_z$ polarization, $a = 1$ m, $z' = 0$, $x' = 0.3$ m, $k_0 a = 50$, $\theta_{bw} = 40°$, $\theta_{elv} = -10°$.

excitation coefficient c_m is real for a real source function without any tilt, and becomes complex if there is a tilted source.

A MATLAB code, *PPlate_Tilt.m*, is prepared for the calculation of the field distribution inside the parallel plate waveguide in terms of mode summation for a tilted Gaussian source. An example for horizontal polarization is shown in Fig. 6.3. Here, fields vs. range inside a 1 m wide plate at 0.3 m is pictured. The source is at 0.3 m height. The number of modes to create Gaussian source at the initial range is 40. Figure 6.3 also displays fields vs. height at two different distances inside the same plate.

6.4 A Hybrid Ray + Image Method

The image method can be used to satisfy an infinite number of reflections with an infinite number of images between two plates. The images are placed with respect to the position of the plates. The unit field (E_y for TE$_z$ case; H_y for TM$_z$ case) can be created by an incident line source at (z', x') as

$$F_i = \frac{\exp(ik_0 r)}{\sqrt{k_0 r}} \qquad (6.12)$$

where $r = \sqrt{(x - x')^2 + (z - z')^2}$. Note that (6.12) is suitable for far-field region. If near-field region is significant, Hankel function can be used instead of exponential function. The first image sources appear at $(z', -x')$ with respect to $x = 0$ plate, and at $(z', 2a - x')$ with respect to $x = a$ plate.

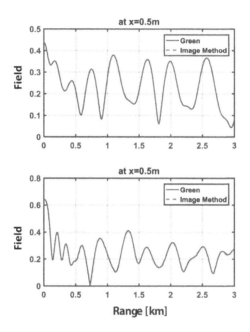

Figure 6.4 (Top) Electric fields vs. range (TE$_z$ case), (bottom) magnetic fields vs. range (TM$_z$ case) with (solid) Green's function (100 modes), (dashed) image method (20 images): $a = 1$ m, $z' = 0$, $x' = 0.4$ m, $k_0 a = 50$.

A short MATLAB script, *PPlate_Image.m*, is for the calculation of the fields inside the parallel plate waveguide in terms of image method. A comparison among mode summation with Green's function and image method for the line-source excitation is done and the results are given in Fig. 6.4 for horizontal and vertical polarizations, respectively. The figures show fields vs. range for specified source and observer heights. Here, 100 modes are used in mode summation and 20 images are used in the image method. As observed, the agreement among the results is very impressive.

6.5 Numerical Models

It is shown in the previous part that analytical reference data can be generated using any of the two models, mode summations or image method.

6.5.1 Parabolic Equation Models: SSPE and FEMPE

The PE wave propagator for a parallel plate waveguide filled with air uses the PE and represents one-way, forward propagation under the paraxial approximation (i.e., near-axial propagation).

The SSPE solution uses a longitudinally marching procedure with the use of DFT. First, an antenna pattern is injected as the initial vertical field profile. Then, this initial field is propagated longitudinally and the transverse field profile at the next range is obtained for $n = 1$

$$u(z + \Delta z, x) = F^{-1} \left\{ C(k_x) F \left\{ u(z, k) \right\} \right\} \tag{6.13}$$

where $C(k) = \exp\left(-ik^2 \Delta z / \left(k_0 + \sqrt{k_0^2 - ck^2}\right)\right)$. Here, F refers to the discrete sine/cosine transform (DST/DCT) while considering (TE_z/TM_z) BCs, and k_x is the spectral variable. The parameter c is 0 (1) for the narrow (wide)-angle model. The new profile $u(z + \Delta z, x)$ is then used in (6.13) as the initial profile and the procedure is repeated; the profile $u(z + 2\Delta z, x)$ at the next range is obtained. The procedure is repeated continuously until the propagator reaches the desired range. A MATLAB module, *PPlate_SSPE.m*, is available for the calculation of fields vs. range or height at a given height or range using the SSPE method.

A FEMPE is based on the division of the vertical profile between lower plate at $x = 0$ and upper plate at $x = a$ into nodes. A vertical antenna pattern is propagated longitudinally using the Crank–Nicolson method. See Section 3.6 for details. A MATLAB module, *PPlate_FEMPE.m*, is available for the calculation of fields vs. range or height at a given height or range using the FEMPE method.

Fields vs. range/height simulated via narrow- and wide-angle SSPE models are shown in Fig. 6.5 for horizontal and vertical polarizations, respectively. A downward tilted Gaussian beam hits the lower plate first and it then bounces back and hits the upper plate. As observed, there is a significant difference between these two plots. In order to see which is accurate, the same fields vs. range/height plot is produced and reference data is generated via the mode summation model for a tilted Gaussian source. As observed, the agreement between wide-angle SSPE and the reference data is impressive. Two vertical cuts are taken from these plots and are shown separately in Fig. 6.6 as fields vs. height at a constant range. These plots show that the agreement between the analytical reference solution and the wide-angle SSPE is better and that there is disagreement with the narrow-angle SSPE.

Assuming a horizontally polarized Gaussian source at 1 GHz with 3 dB beamwidth of $7°$, located at mid-height and tilted $30°$ downward, inside a 8 m wide parallel plate PEC waveguide, the performances of narrow- and wide-angle PE tools with respect to the analytical result are illustrated in Fig. 6.7 by means of 3D field maps. The

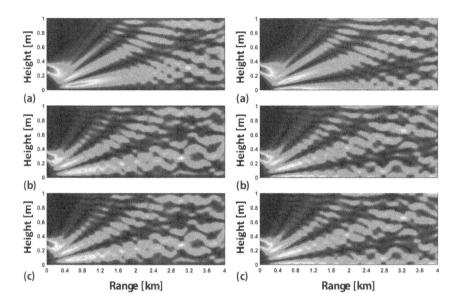

Figure 6.5   Fields vs. range/height for a tilted Gaussian beam, (a) narrow SSPE, (b) wide SSPE, (c) Green's function (41 modes): (Left) TE_z case (horizontal polarization), (right) TM_z case (vertical polarization), $a = 1$ m, $z' = 0$, $x' = 0.3$ m, $k_0 a = 50$, $\theta_{bw} = 40°$, $\theta_{elv} = -15°$.

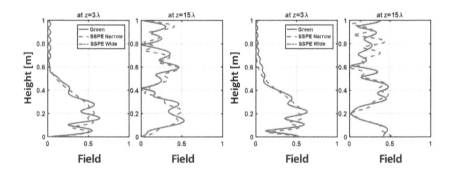

Figure 6.6   Fields vs. height for a tilted Gaussian beam, (left) horizontal polarization, (right) vertical polarization, $a = 1$ m, $z' = 0$, $x' = 0.3$ m, $k_0 a = 50$, $\theta_{bw} = 40°$, $\theta_{elv} = -15°$ at $z = 3\lambda$, and at $z = 15\lambda$, (solid) Green's function (41 modes), (dashed) narrow SSPE, (dashed-dotted) wide SSPE.

two field maps generated with the narrow-angle FEMPE and SSPE tools seem to be logical and physical. A down-tilted Gaussian beam is bouncing up and down while longitudinally propagating with the interference patterns exactly alike shown there. On the other hand, the true 3D wave patterns are shown on the right side of Fig. 6.7. The analytical exact solution is exactly the same as the FEMPE and SSPE results

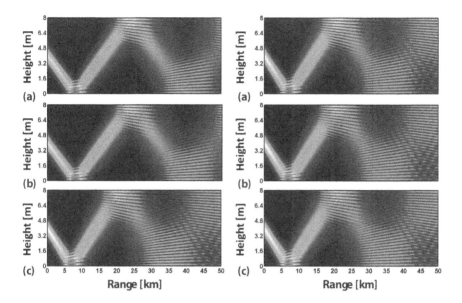

Figure 6.7   Electric fields vs. range/height for a tilted Gaussian beam, (left) narrow case, (right) wide case, (a) FEMPE, (b) SSPE, (c) Green's function (49 modes): TE$_z$ polarization, $a = 8$ m, $z' = 0$, $x' = 4$ m, $k_0 a = 168$, $\theta_{bw} = 7°$, $\theta_{elv} = -30°$.

and is not shown here. Hence, these results validate the accuracy of the wide-angle FEMPE and SSPE, and show that the narrow-angle FEMPE and SSPE cannot handle large tilt angles. To validate the narrow-angle model, the angle must be constrained to be less than $10°$–$15°$ while considering both beamwidth and tilt angles. Figure 6.8 shows vertical field profiles computed via analytical, FEMPE, and SSPE models in a highly oscillatory region at $z = 100\lambda$. The results computed via narrow-angle FEMPE and SSPE tools are compared against the reference solution. As observed, FEMPE and SSPE agree very well but they both disagree with the reference (analytical) solution. This is an interesting example since it shows that both FEMPE and SSPE pass the verification test but fail the (model) validity test.

6.5.2   Method of Moments

The MoM [84] is one of the oldest numerical EM methods. In this method, a discrete model of the object under investigation is first created from small (compared to wavelength) pieces, called segments or patches, and then an $N \times N$ system of equations is built with N unknown segment/patch currents, N known segment voltages calculated from Green's function of the problem, and known (analytically, but requires numerical computation) $N \times N$ segment/patch impedances (it requires N^3 operations). The model is closed form and stable but necessitates large memory and high-speed computers especially for high-frequency applications. Also, it requires the derivation of Green's function of the problem.

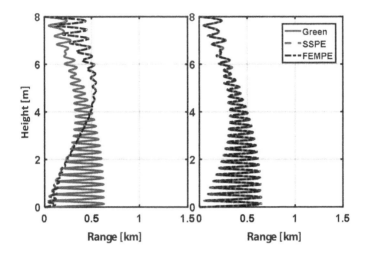

Figure 6.8 Electric fields vs. height at $z = 100\lambda$ for a tilted Gaussian beam, TE_z polarization, $a = 8$ m, $z' = 0$, $x' = 4$ m, $k_0 a = 168$, $\theta_{bw} = 7°$, $\theta_{elv} = -30°$ (left) narrow case, (right) wide case, (solid) Green's function (49 modes), (dashed) SSPE, (dashed-dotted) FEMPE.

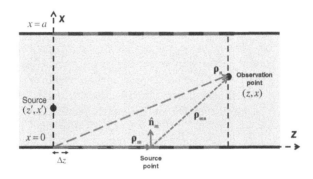

Figure 6.9 A parallel plate waveguide and the source/observer for the MoM (ρ_m: source segment, ρ_n: observer segment).

The MoM can be used to find the propagation of horizontally (TE_z case) and vertically (TM_z case) polarized waves by using the electric field integral equation and the magnetic field integral equation, respectively. Unfortunately, the MoM suffers from resonances in waveguiding structures [94], therefore its direct application is a challenge. Recently, a novel multiple MoM model (Mi-MoM) is introduced for this purpose [46]. In this method, the integral equation is converted to the corresponding matrix equation via the discretization of the plates at $x = 0$ and $x = a$ (see Fig. 6.9). The lengths of the line segments on the plates are chosen to be less than one-tenth of the wavelength to satisfy the surface current constant at each line segment. Then, an

$N \times N$ system of equations is constructed with the unknowns of N segment currents and the system is solved numerically.

The point-matching MoM technique [84,95] with the closed form matrix equation $[V] = [Z][I]$ where $[I]$ contains the unknown segment currents, $[V]$ corresponds to excitation voltage matrix using the incident field evaluated at the midpoints of each line segment, $[Z]$ is the $N \times N$ impedance matrix for the PEC plate. Using segment currents, the scattered fields are obtained. Finally, fields at a specified observation point are calculated in terms of the incident field and scattered fields caused by the source-induced segment currents on plates.

The Mi-MoM procedure for propagation modeling inside resonating structures is based on taking into account multiple induced segment currents and may be outlined as follows [46]:

- First, discretize the upper and lower boundaries. Use $N/2$ segments for the lower boundary and $N/2$ segments for the upper boundary. Label all segments from 1 to N.

- Calculate segment currents $[I]$ from $[I] = [Z]^{-1}[V]$ and calculate the scattered and total fields using either E_y^{inc} or H_y^{inc} for TE$_z$ and TM$_z$ polarizations, respectively.

- For a given source point, calculate the distances to all segments and all segment voltages, using either E_y^{inc} or H_y^{inc} for TE$_z$ and TM$_z$ polarizations, respectively. This will yield $[V]$.

- Calculate the impedance matrix Z_{mn}.

- The segment currents induced by the external source on the upper plate excite fields on segments on the lower plate and vice versa. For the first segment of the lower plate, calculate the distances to all segments on the upper plate and the segment voltages, using either E_y^{inc} or H_y^{inc} for the TE$_z$ and TM$_z$ polarizations, respectively. Repeat this for all segments on the lower plate and find the voltages on the upper plate caused by the segments on the lower plate.

- Do the same for the segments on the upper plate and find the voltages on the lower plate caused by the segments on the upper plate. This will yield second round $[V]$.

- Use the same impedance matrix Z_{mn} and calculate second round segment currents $[I]$ from $[I] = [Z]^{-1}[V]$ and scattered and total fields for TE$_z$ and TM$_z$ polarizations, respectively.

- Repeat the procedure and find third round segment currents and scattered and total fields caused by these currents.

- Repeat the whole procedure until the desired accuracy is reached.

An alternative way is to find the first round segment currents and then use the image method. First, all segment currents of the upper and lower plates are obtained.

Then, the boundaries are removed and image segments are added with respect to the upper and lower plates. Finally, the field contributions from the currents of the segments and image segments are superposed at the receiver. A short MATLAB script, *PPlate_MoM.m*, for the Mi-MoM calculation of fields vs. range/height at a given height/range point is prepared. The classical MoM can also be used together with the IM in hybrid form. This is achieved by finding segment currents once and then using images of all segments on both sides of the boundaries. A short MATLAB script, *PPlate_MoM2.m*, for the hybrid MoM+IM calculation of fields vs. range/height at a given height/range is also available.

An example of the Mi-MoM procedure with 40 iterations is given in Fig. 6.10. The result is compared to mode summation with 42 modes. As shown, very good agreement is obtained.

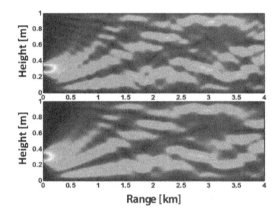

Figure 6.10 Fields vs. range/height (TE_z case): (top) mode sum with 42 modes, (bottom) Mi-MoM with 40 iterations: $a = 1$ m, $z' = 0$, $x' = 0.3$ m, $k_0 a = 50$, $dz = dx = 0.01$ m, $\theta_{bw} = 45°$, no tilt.

CHAPTER 7

WAVE PROPAGATION INSIDE THREE-DIMENSIONAL RECTANGULAR WAVEGUIDE

7.1 Introduction

Parabolic equation models have been widely used in solving both underwater acoustic and tropospheric EM wave propagation problems since its first introduction of the PE method [18]. The PE models one-way, forward EM propagation over a non-flat, lossy Earth's surface through a non-homogeneous atmosphere. It is simple to understand, code, and apply, and highly effective in complex wave environments. It is impossible to include all contributions in this area; the list would always be incomplete. The book by Levy [5] might be a good source for the PE model and early studies on propagation modeling. A few simple but highly effective propagation virtual tools have also been introduced [10, 12, 96]. There are also two-way PE models [2, 27, 43, 67, 68], which have been effectively used in propagation problems where backward scattering is non-negligible.

The growth of mobile communication systems in recent years necessitates realistic 3D propagation modeling in urban regions as well as in tunnels. The PE method has also been used for this purpose. The 3D calibration of various PE models/codes is therefore essential before their applications in complex realistic environments (as done in two dimensions, see [2, 27, 31]). The calibration should be

Radio Wave Propagation and Parabolic Equation Modeling, First Edition. By Gökhan Apaydin, Levent Sevgi
© 2017 by the Institute of Electrical and Electronic Engineers, Inc. Published 2017 by John Wiley & Sons, Inc.

performed against available analytically exact models, which should be accurately numerically computed. The classical rectangular waveguide problem is chosen for this purpose. Rectangular and/or circular waveguides may also be chosen, but this structure is chosen to eliminate the discretization errors (in fact, the waveguide may have an arbitrary cross-section in the PE models, but in that case, an analytical reference solution is not available). Propagation inside this waveguide can be exactly modeled in terms of modal summation and any given source distribution can be analytically represented as accurately as desired. Source tilt can also be included which makes modal excitation coefficients complex.

In this chapter, the 3D SSPE, FEMPE, and alternate direction implicit parabolic equation (ADIPE) algorithms are developed, tested against analytically exact data, and calibrated. The 3D rectangular waveguide is used as the canonical structure and its solutions are used as the reference.

7.2   Three-Dimensional Rectangular Waveguide Model

Consider 3D propagation inside a rectangular cross-section waveguide with nonpenetrable boundaries, located longitudinally along the z-direction (see Fig. 7.1).

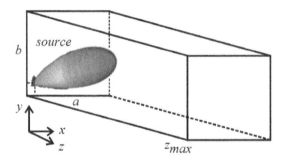

Figure 7.1   The 3D rectangular waveguide, the coordinates, and a tilted source distribution.

This waveguide acts as a high-pass filter and the cutoff frequency is determined by the width (a) and height (b) of the waveguide ($0 < x < a, 0 < y < b$). The rectangular waveguide is a canonical EM structure and both Green's function (source-driven) and eigenvalue (source-free) solutions for the TE and TM cases are well established.

Any given initial source distribution inside a rectangular waveguide can be represented in terms of a modal summation as [97]

$$\psi(x, y, z) = \sum_{p=p_1}^{N_1} \sum_{q=q_1}^{N_2} c_{pq} B\left(\frac{p\pi}{a}x\right) B\left(\frac{q\pi}{b}y\right) \exp(i\beta_{pq}z) \qquad (7.1)$$

where $\beta_{pq} = \sqrt{k_0^2 - (p\pi/a)^2 - (q\pi/b)^2}$ is the longitudinal propagation constant, N_1 and N_2 are the total number of modes of the directions x and y; k_0 is the free-space wavenumber, and c_{pq} is the modal excitation coefficient derived from the orthonormality condition as

$$c_{pq} = v_{pq} \int_{x=0}^{a} \int_{y=0}^{b} u_{in}(x,y) B\left(\frac{p\pi}{a}x\right) B\left(\frac{q\pi}{b}y\right) dxdy \qquad (7.2)$$

v_{pq} is the normalization constant

$$v_{pq} = \left(\int_{x=0}^{a} \int_{y=0}^{b} B^2\left(\frac{p\pi}{a}x\right) B^2\left(\frac{q\pi}{b}y\right) dxdy \right)^{-1} \qquad (7.3)$$

and $u_{in}(x,y)$ is the initial field profile, which is generated from a Gaussian antenna pattern

$$
\begin{aligned}
u_{in}(x,y) \;&= \exp\left[ik_0 x \sin\theta_{elvx} - \frac{(k_0 \sin(\theta_{bwx}/2)(x-x_s))^2}{2\ln 2} \right] \\
&\times \exp\left[ik_0 y \sin\theta_{elvy} - \frac{(k_0 \sin(\theta_{bwy}/2)(y-y_s))^2}{2\ln 2} \right]
\end{aligned}
\qquad (7.4)
$$

specified by its position (x_s, y_s), beamwidth $(\theta_{bwx,bwy})$, and tilt (elevation) angle $(\theta_{elvx,elvy})$. Note that B shows either a sine or a cosine function starting from either $p_1 = q_1 = 1$ or $p_1 = 1, 0; q_1 = 0, 1$ for horizontal and vertical polarizations, respectively.

A short MATLAB script, *Rect_Waveguide.m*, that calculates field variations inside a 3D rectangular waveguide using the modal summation approach for a specified accuracy, is written to serve as the analytical reference data.

7.3 Three-Dimensional Parabolic Equation Models

The 3D PE wave propagator for rectangular waveguides filled with air uses the standard PE

$$\left(\frac{\partial^2}{\partial x^2} + \frac{\partial^2}{\partial y^2} + 2ik_0 \frac{\partial}{\partial z} \right) u(x,y,z) = 0 \qquad (7.5)$$

representing one-way, forward propagation under the paraxial approximation (i.e., near-axial propagation). Here, x and y stand for the transverse coordinates, and z is the longitudinal coordinate. Since the direction of wave propagation is predominantly along the z-axis, the reduced function $u(x,y,z)$ is used by separating the rapidly varying phase term from $\psi(x,y,z) = \exp(ik_0 z)u(x,y,z)$ which corresponds to either electric or magnetic fields for the horizontal (TM) and vertical (TE) polarizations, respectively.

7.3.1 SSPE Model

The standard SSPE uses a longitudinally marching procedure [5] with the use of the DFT. First, an antenna pattern representing the initial 2D field profile is injected (the source profile may be given in either the spatial or transverse wavenumber domains).

This initial field, given in (7.4), is then longitudinally propagated, and the transverse field profile at the next range is obtained. This new profile is then used as the initial profile for the next step and the procedure is repeated until the propagator reaches the desired range given as [97]

$$u(x, y, z + \Delta z) = F^{-1} \{C(k_x)C(k_y)F\{u(x, y, z)\}\} \tag{7.6}$$

where $C(k) = \exp\left(-ik^2\Delta z / \left(k_0 + \sqrt{k_0^2 - con \times k^2}\right)\right)$, F refers to DST/DCT while considering BCs, k_x and k_y are the spectral variables. The parameter con is either 0 or 1 for the narrow- and wide-angle cases, respectively. The 2D transverse fields are obtained along z at each range step of Δz. Dirichlet BC and Neumann BC (DBC/NBC) are considered for horizontal and vertical polarizations, respectively (for the PEC case) [31].

7.3.2   FEMPE Model

The idea of FEMPE model is first to divide the 2D transverse-domain into subdomains. Using the initial field profile, which is generated from a Gaussian antenna pattern given in (7.4); the approximated field values at discrete nodes in domain are

$$u_{ap}(x, y, z) = \sum_{e=1}^{N} \sum_{j=1}^{2} c_j^e(x, y)B_j^e(z) \tag{7.7}$$

where N is the number of elements, $c_j^e(x, y)$ indicates the coefficients of the unknown function, and $B_j^e(z)$ are the piecewise linear Lagrange basis functions. The approximated fields are then longitudinally propagated by the application of Crank–Nicolson approach, which is based on the improved Euler method with DBC and NBC at each range for horizontal and vertical polarizations, respectively (PEC case) [31,44,97].

7.3.3   ADIPE Model

The ADIPE is a finite-difference method introduced for the solution of 2D or 3D complex EM propagation problems with the following second-order derivative and central-difference approximation [97,98]

$$u_{m,l}^n + r_x \left(u_{m+1,l}^n - 2u_{m,l}^n + u_{m-1,l}^n\right) + r_y \left(u_{m,l+1}^n - 2u_{m,l}^n + u_{m,l-1}^n\right)$$

$$= u_{m,l}^{n+1} - r_x \left(u_{m+1,l}^{n+1} - 2u_{m,l}^{n+1} + u_{m-1,l}^{n+1}\right) - r_y \left(u_{m,l+1}^{n+1} - 2u_{m,l}^{n+1} + u_{m,l-1}^{n+1}\right) \tag{7.8}$$

where $r_x = \Delta z / \left(4ik_0\Delta x^2\right)$, $r_y = \Delta z / \left(4ik_0\Delta y^2\right)$. The equations have a relatively simpler structure and two 1D problems are solved at each range step for the 2D PE. First, one direction is kept fixed; and then the other. In this way, both equations can be written in tridiagonal matrix forms. The 2D ADIPE model [98] uses the following difference equations for $1 < n < N_z$

$$-r_y \left(u_{m,l-1}^{n+1/2} + u_{m,l+1}^{n+1/2} \right) + (1 + 2r_y) u_{m,l}^{n+1/2} = r_x \left(u_{m-1,l}^{n} + u_{m+1,l}^{n} \right) + (1 - 2r_x) u_{m,l}^{n}, \text{ for } 1 < m < N_x \qquad (7.9)$$

$$-r_x \left(u_{m-1,l}^{n+1} + u_{m+1,l}^{n+1} \right) + (1 + 2r_x) u_{m,l}^{n+1} = r_y \left(u_{m,l-1}^{n+1/2} + u_{m,l+1}^{n+1/2} \right) + (1 - 2r_y) u_{m,l}^{n+1/2}, \text{ for } 1 < l < N_y \qquad (7.10)$$

where N_x, N_y, N_z are the node numbers on the x-, y-, and z-axes. The ADIPE model is easier to solve than the traditional Crank–Nicolson approach.

7.4 Tests and Calibration

Field distributions inside square, rectangular, and circular cross-section waveguides are well known. All PE-based techniques (SSPE, FEMPE, and ADIPE) manage to give cross-sectional field distributions as long as the operating frequency is good above the cutoff frequency of the dominant mode. Consider the dominant mode (TE_{10}) inside a rectangular waveguide; the transverse field distribution is *half-sine* as illustrated in Fig. 7.2. The *sine* function represents a standing wave and can be decomposed into two exponential functions (i.e., two transverse traveling waves). The wave vectors of these traveling waves are shown with thick black block arrows. Three cases are shown in Fig. 7.2. The wave vectors are almost vertical (horizontal) if the frequency is very close to (far from) the cutoff frequency. The red arrows show the validity region (angular spectrum) of the PE models. The angular validity range is approximately $\pm 15°$ to $20°$ for narrow-angle PE models and $\pm 40°$ to $45°$ for wide-angle PE models, therefore any transverse field contribution with wave vectors beyond these angles cannot be handled with the PE models.

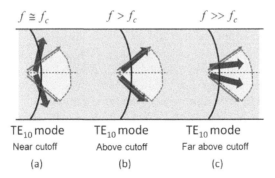

Figure 7.2 The transverse field distribution of the dominant TE_{10} mode and the two wave vectors for three different frequency ranges (f_c is the cutoff frequency).

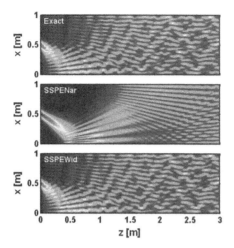

Figure 7.3   Three-dimensional color plots of fields vs. range–height variations inside a parallel plate PEC waveguide: (top) exact, (middle) SSPE narrow-angle propagator, (bottom) SSPE wide-angle propagator ($\Delta x = 2.5$ mm, $\Delta z = 5$ mm) [96].

The performance of narrow- and wide-angle SSPE tools with respect to the analytical result is illustrated in Fig. 7.3 [96]. Here, a 1 m wide parallel PEC plate waveguide is taken into account. A horizontally polarized Gaussian beam, tilted down 45° at 3 GHz with 3 dB beamwidth of 30° is excited by a source located at the mid-height of the waveguide. As observed, the wide-angle SSPE agrees very well with the analytically exact (reference) solution. On the other hand, while it visually seems to be physical and correct, the narrow-angle SSPE is incorrect.

A square waveguide with dimensions of 4 m × 4 m, having a cutoff frequency of 37.5 MHz is considered in Fig. 7.4. These dimensions are selected as appropriate for the size of a typical tunnel. The frequency is 3 GHz, which means the propagation inside the waveguide is good above the cutoff frequency. Horizontally polarized waves (TM) are taken into account. The transverse discretization values are $\Delta x = 0.8\lambda$ and $\Delta y = 0.8\lambda$, that means the waveguide is modeled using 50 × 50 cells in the transverse-domain. The longitudinal step is $\Delta z = 10\lambda$. A 2D, untilted, centrally located Gaussian spatial source is used with a beamwidth of $\theta_{bw} = 4.34°$. This untilted source is represented in terms of modes of the structure and the number of modes in the transverse-domain is 19 in the x and y directions, within a relative maximum error of 1e-7. The relative maximum error is defined as the maximum difference in the absolute values of the Gaussian field and the modal solution divided by the maximum absolute value of the Gaussian field. The modal solution is assumed as the reference.

Propagation inside the waveguide is computed with the SSPE, FEMPE, and ADIPE simulators. Figure 7.4 shows the cross-sectional field distributions of the untilted and tilted sources at three different ranges. The rows correspond to the reference, SSPE,

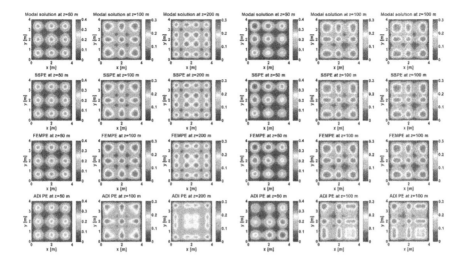

Figure 7.4 Two-dimensional cross-sectional field distributions of untilted and tilted sources: From top to bottom, the rows are for the reference, SSPE, FEMPE, and ADIPE simulations. From left to right, the columns are for 50 m, 100 m, and 200 m ranges, respectively. Source position: $(x_s, y_s, z_s) = (20\lambda, 20\lambda, 0)$; $f = 3$ GHz, $\theta_{bw} = 4.34°$, (left) $\theta_{elvx,elvy} = (0, 0)$, (right) $\theta_{elvx,elvy} = (-0.5°, 0.5°)$, TM Polarization, $\Delta x = \Delta y = 0.8\lambda$, $\Delta z = 10\lambda$.

FEMPE, and ADIPE solutions; the columns belong to ranges of 50 m, 100 m, and 200 m, respectively. According to the results in these figures, the SSPE and FEMPE are more accurate than the ADIPE for the same discretization parameters [97].

When the source position is not centrally located inside the waveguide, the wide-angle SSPE model gives the best approximation among the models/simulators. Figure 7.5 shows 2D cross-sectional field distribution for the source position $(x_s, y_s) = (20\lambda, 20\lambda)$ and tilt angles of $\theta_{elvx,elvy} = (3°, 4°)$ by using the wide-angle SSPE model. At this time, the number of modes in the transverse-domain is $(N_1, N_2) = (47, 49)$ in the x and y directions, respectively, within a relative maximum error of 1e-7. Figure 7.5 also shows field distribution on a 2D plane cut longitudinally (i.e., on the zy-plane) at $x = 20\lambda$ for the set of parameters, obtained by using the modal solution and the wide-angle SSPE model. As observed, the agreement between the two models is impressive.

Two-dimensional field distributions for different ranges are mostly good for visualization purposes and are not good for precise comparisons. Comparisons are performed by using either 1D field variations or the root mean square (rms) error with respect to the modal solution. The rms error is defined as the square root of the sum of the squares of the differences between the approximated field and the modal solution, divided by the number of nodes ($N_x \times N_y$). Longitudinal field variations at the center of the waveguide computed with four different approaches are plotted in Fig. 7.6. As observed, there is a very good agreement among the reference, SSPE, and FEMPE solutions. The ADIPE solution also agrees with the reference data,

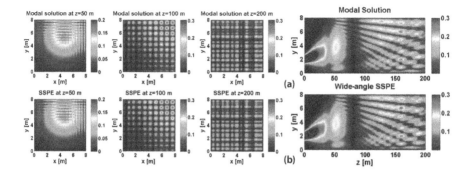

Figure 7.5   Two-dimensional cross-sectional field distributions. (Left) The first row is the reference solution and the second row is the wide-angle SSPE simulation. From left to right, the columns are 50 m, 100 m, and 200 m ranges. (Right) Fields at $x = 20\lambda$. Source position is $x_s = 20\lambda$, $y_s = 20\lambda$, $z_s = 0$; $f = 3$ GHz, $\theta_{bw} = 4.34°$, $\theta_{elvx,elvy} = (3°, 4°)$, TM polarization, $\Delta x = 0.8\lambda$, $\Delta y = 0.8\lambda$, $\Delta z = 10\lambda$. The rms error at 50 m is 0.96%; at 100 m, 3.80%; at 200 m, 7.44%.

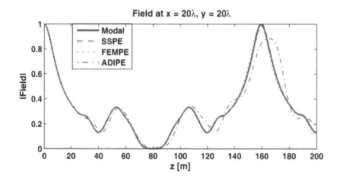

Figure 7.6   Longitudinal field variations at the center of the rectangular waveguide. Source position is $x_s = 20\lambda$, $y_s = 20\lambda$, $z_s = 0$; $f = 3$ GHz, $\theta_{bw} = 4.34°$; TM polarization; $\Delta x = 0.8\lambda$, $\Delta y = 0.8\lambda$, $\Delta z = 10\lambda$.

but not as well as the other two solutions for the set of discrete parameters used. Moreover, transverse field variations at $y = 20\lambda$ of an untilted source are taken into consideration. All numerical methods are compared with the modal solutions at three different ranges. As observed, the SSPE and FEMPE models agree very well with the reference solution. This agreement is obtained in Fig. 7.7, which shows the vertical field profile with respect to the x-axis at the $z = 0$, $z = 100$ m, and $z = 200$ m ranges [97].

Moreover, standard PE models cannot show the effects of the interaction between the forward and backward waves, especially if there are obstacles inside waveguides.

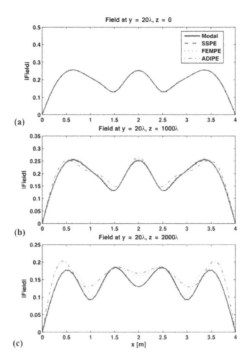

Figure 7.7   Transverse field variations ($y = 20\lambda$) of an untilted source at different ranges: (a) $z = 0$ m, (b) $z = 100$ m, (c) $z = 200$ m using the reference, SSPE, FEMPE, and ADIPE simulations; source position is $x_s = 20\lambda$, $y_s = 20\lambda$, $z_s = 0$; $f = 3$ GHz, $\theta_{bw} = 4.34°$; $\theta_{elvx,elvy} = (0,0)$; TM polarization; $\Delta x = 0.8\lambda$, $\Delta y = 0.8\lambda$, $\Delta z = 10\lambda$.

In the presence of such obstacles, the effects of forward and backward waves must be considered in order to get more accurate results [93]. At this time, the forward PE model computes as the previous model. Then, the total field is obtained by superposing the forward and backward fields. The backward fields can be found by taking the tangential fields of the obstacle to be zero while considering the same algorithm of the forward PE model. Figure 7.8 shows 2D cross-sectional field distributions for various ranges inside a rectangular waveguide terminated at 200 m range. As observed, good agreement is obtained by using the two-way 3D PE models. Note that the fastest method is the SSPE; roughly speaking, the ADIPE simulations take twice the time of the SSPE simulations, and the FEMPE is the slowest method (nearly a factor of two slower for the sets of parameters used in the simulations). This is because

- The SSPE uses the 2D DST (TM) or DCT (TE) at each range step.

- The ADIPE solves two tridiagonal matrices (for the x and y directions) at each range step.

- The FEMPE solves $N \times N$ matrices ($N = N_x \times N_y$) at each range step.

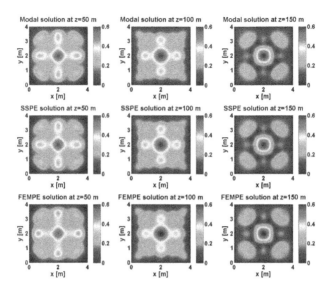

Figure 7.8   Two-dimensional cross-sectional field distributions inside a rectangular waveguide terminated at 200 m range. From top to bottom, the rows show the reference solution, SSPE, and FEMPE simulations. From left to right, the columns are 50 m, 100 m, and 150 m ranges. Source position is $x_s = 20\lambda$, $y_s = 20\lambda$, $z_s = 0$; $f = 3$ GHz, $\theta_{bw} = 4.34°$; $\theta_{elvx,elvy} = (0,0)$; TM polarization; $\Delta x = 0.8\lambda$, $\Delta y = 0.8\lambda$, $\Delta z = 10\lambda$.

- For PEC boundaries, both the ADIPE and FEMPE matrices are range independent, therefore they can be calculated once.

- The SSPE range steps can be chosen to be much larger than the ADIPE and FEMPE.

- The range step sizes of the ADIPE and FEMPE are similar because they are both based longitudinally on the finite-difference approach.

- The transverse discretization is similar for all three methods.

It should be noted that the priorities would change in realistic tunnel environments with lossy (CBC) irregular walls and non-flat bottom irregularities, therefore one may suggest the use of

- The SSPE for PEC walls and irregularities; this would be the fastest and most accurate.

- The ADIPE and FEMPE for CBC walls and irregularities; ADIPE would be faster, but FEMPE would be more accurate. They both would be a few orders slower than the SSPE.

- The SSPE with the DMFT for CBC walls and irregularities.

CHAPTER 8

TWO-WAY PE MODELS

8.1 Formulation of Two Way FEMPE Method

The standard PE method is a one-way, forward propagation model, because of ignoring the backward propagation term. The standard PE model cannot reflect the effect of the interaction between the forward and backward waves, especially if there are valleys or hills with steep slopes along the propagation path. In the presence of such obstacles, the effects of not only forward but also backward-reflected, refracted, and diffracted waves must be very well predicted to be able to get reliable results.

The algorithm can be applied to a variable terrain by using staircase approximations, as illustrated in Fig. 8.1. Two-way PE algorithm is as follows:

- Step 1: The fields inside terrain are set to zero for staircase approximation. If the vertical field meets a vertical terrain facet, it is split into two components propagating in forward and backward directions.

- Step 2: The forward field continues in the usual way after setting it to zero on the vertical terrain facet (see Fig. 8.1). In other words, the fields (at $z + \Delta z$) obtained from the previous vertical field (at z) are set to zero inside the terrain between $x = 0$ and $x = xter(z + \Delta z)$.

Radio Wave Propagation and Parabolic Equation Modeling, First Edition. By Gökhan Apaydin, Levent Sevgi
© 2017 by the Institute of Electrical and Electronic Engineers, Inc. Published 2017 by John Wiley & Sons, Inc.

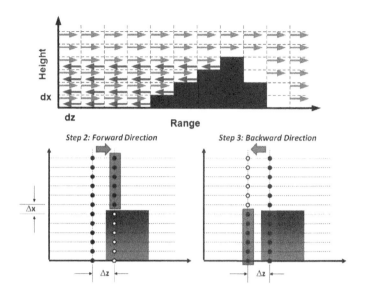

Figure 8.1 (Top) Two-way PE propagation model with forward (right arrows) and backward (left arrows) propagating waves generated from terrain reflections, (bottom) the vertical fields split into forward and backward fields.

- Step 3: The field must be partially reflected from the terrain facet, therefore first the initial field of the backward field is obtained by imposing the BCs at the facet (i.e., the tangential field must be zero on the PEC facet), and then this initial field is marched back in the $-z$ direction by reversing the signs of k_0 and Δz. Namely, the backward-propagating waves are initiated from the waves between $x = xter(z)$ and $x = xter(z + \Delta z)$, and then, propagated in the reverse direction (see Fig. 8.1).

- Step 4: The same form of the PE is derived, as expected, for the reduced function in the backward propagation, but the original field is expressed as $\psi_b(z, x) = u_b(z, x) \exp(-ik_0 z)$.

The two-way FEMPE (2W-FEMPE) approach is the iterative implementation of the one-way FEMPE (1W-FEMPE) algorithm by simply rotating the direction of propagation in a forward–backward manner to estimate the multiple-reflection effects.

Both forward and backward fields continue to march out in their own propagation directions. Each time the wave hits a terrain facet, it is again split into forward and backward components. The total field inside the domain is then determined by the superposition of the backward and forward fields at each range step. It is useful to note that the convergence of the algorithm is achieved because, as the iterations are carried out, the field contributions of the multiple reflections decrease with regard to the $1/\sqrt{r}$ term in 2D Green's function. The convergence of the algorithm is checked against a certain threshold criterion, which compares the total fields at each iteration.

8.2   Formulation of Two Way SSPE Method

The two-way SSPE (2W-SSPE) algorithm is basically the iterative implementation of the one-way SSPE (1W-SSPE) by simply switching the direction of propagation in a forward–backward manner to estimate the multiple-reflection effects. During forward propagation, the split-step solution is employed to march the solution. However, it is evident with regard to the physics of the problem, the field must be partially reflected from the terrain facet. As 2W-FEMPE, this is achieved in the two-way algorithm in such a way that first the initial field of the backward field is obtained by imposing the BCs at the facet (i.e., the tangential field must be zero on the PEC facet), and then this initial field is marched back in the $-z$ direction by reversing the signs of k_0 and Δz.

8.3   Flat Earth with Infinite Wall

Calibration tests are performed both 2W-FEMPE and 2W-SSPE codes over flat Earth with the infinite wall against data generated via image method as well as GO+UTD solutions. In all scenarios, the frequency is set to 3 GHz.

A MATLAB code, *twowaype.m*, is available for the calculation of PF over a PEC ground with a vertically infinite-extend wall. Here, 2W-SSPE, 2W-FEMPE, and image methods are taken into consideration. A vertically infinite-extent wall at range 40 km is described over the flat Earth with PEC boundaries in Fig. 8.2. The source is at 50 m. Excellent agreement among the results shows the effectiveness of both 2W-SSPE and 2W-FEMPE algorithms. Also note that FEMPE necessitates smaller range steps as compared with the SSPE method. The SSPE requires smaller height steps for vertical polarization as compared to the horizontal polarization calculations. On the other hand, FEMPE gives the same accuracy with larger height step size for the vertical polarization, but still slower than the SSPE because of longitudinal marching with much smaller range steps.

8.4   Flat Earth with Single and Double Knife Edges

The 2W-FEMPE model is validated and verified against analytical approximate models as well as the 2W-SSPE model. In the first two scenarios, the tests are performed against the GO+UTD results, assuming that the frequency is 3 GHz and the polarization is horizontal, therefore DBC has been taken into consideration at the surface. It is worthwhile to note that the calibration by GO+UTD is possible only at high frequencies where the GO interpretation is valid. The first scenario belongs to a single PEC wall of finite height (i.e., a knife edge), illuminated by a line source over

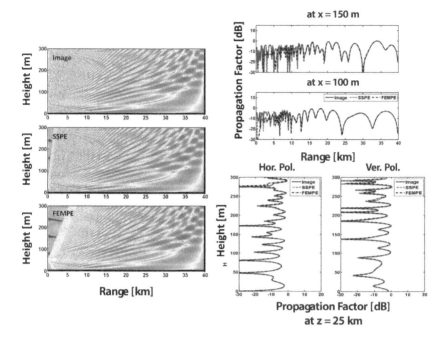

Figure 8.2 (Left/right-top) PF vs. range/height with horizontal polarization, (right-bottom) fields vs. height at $z = 25$ km with different polarizations. (SSPE: $\Delta z = 100$ m, FEMPE: $\Delta z = 12.5$ m).

a PEC ground and in the free space. The approximate solution of this well known and canonical problem in the field of diffraction theory is available by combining the GO technique with some special diffraction methods such as non-uniform GTD [99], UTD [78], PTD [79]. In the GO method, the reflected waves from the ground and the wall are determined by employing the principles of image theory.

Figure 8.3 presents some of the possible GO+UTD components that must be taken into account in the GO+UTD computations. Note that double diffractions have not been included in the GO+UTD calculations. The total field is determined by summing the direct ray, the reflected rays emanating from image sources, and the rays diffracted from the tip of the wall. Note also that these ray contributions are added to the total field only if the LOS condition between each source and the observation point is satisfied.

The first simulation results are presented in Fig. 8.4, assuming that the line source is 250 m above the ground at range $z_s = 0$. The height of the wall is 150 m, the wall is positioned at 40 km range, and the polarization is horizontal, therefore DBC is satisfied at the surface of the wall and the ground. The region between the source and the wall is within the interference region where forward-, ground-, and backward-reflected waves interfere. Beyond the wall, only diffracted waves appear for heights below the source height (i.e., in the shadow region), and forward propagated and

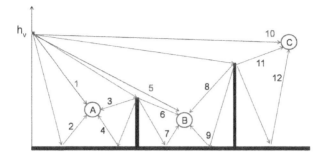

Figure 8.3   GO+UTD modeling and some possible direct, reflected, and diffracted field components.

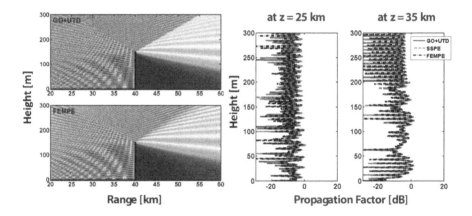

Figure 8.4   PF vs. range/height over flat Earth with single knife edge (150 m height-wall at 40 km range) with horizontal polarization: (left-top) GO+UTD, (left-bottom) FEMPE, (right) PF vs. height at two specified ranges; $f = 3$ GHz, $\Delta x = 0.33$ m, $\Delta z = 12.5$ m.

diffracted waves exist if the receiver height is above the source height. The LOS and reflection boundaries and transitional regions around these boundaries are clearly observed in both maps. Note that in the 2W-FEMPE implementation, the field bounces from the wall only once. Namely, only a single forward–backward field pair is generated without resorting to any iterative scheme. Two vertical slices at 25 km and 35 km (in the interference region) extracted from this 3D field map are presented. The result of the recently introduced 2W-SSPE model is also included in this figure. As observed, an excellent agreement has been obtained from all three models.

The second scenario deals with two walls, one of which is finite and the other one is infinite in height. The finite-height (150 m) wall is located at 40 km, and the infinite-height wall is at 60 km. Figure 8.5 presents the test results (as a 3D field map

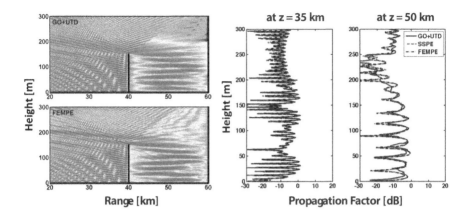

Figure 8.5   PF vs. range/height over flat Earth with two edges (150 m height-wall at 40 km range and an infinite-height wall at 60 km) with horizontal polarization: (left-top) GO+UTD, (left-bottom) FEMPE, (right) PF vs. height at two specified ranges; $f = 3$ GHz, $\Delta x = 0.33$ m, $\Delta z = 12.5$ m.

vs. range/height). The line source is at 250 m above the ground, and the polarization is horizontal. Same as before, DBC is used at the surface of the walls and the ground. Note that the region between the walls exhibits multiple reflections and diffractions from the walls and/or ground. In the 2W-FEMPE realization, the field in this region bounces from both of the walls several times until the contribution becomes negligible (for a given accuracy limit). PF vs. height at 35 km and 50 km ranges of this scenario is plotted. At 35 km range, forward-propagating waves interfere with the backward-reflected waves (from the knife edge wall) and edge-diffracted waves (from the top of the knife edge wall). Since the forward-propagated and backward-reflected contributions are stronger than the edge-diffracted contributions, excellent agreement among the three models is obtained. It is observed that multiple reflections (almost resonance behavior) occur in the region between the walls. It is known that the GTD is a HFA method; whereas the FEMPE and SSPE approaches can account for the diffraction effects up to a certain extent. In comparing the 2W-FEMPE with the GO+UTD approach, the contribution of the waves hitting the walls up to three times is superposed. To have fair comparisons up to third degree of reflections, the GO+UTD code is developed to account for 35 types of rays bouncing from the walls and the ground, in accordance with the LOS criteria. These rays include the reflected waves up to a third degree, as well as the diffracted waves from the finite-height wall. Excellent agreement among the results verifies the performance of the 2W-FEMPE with respect to GO+UTD approach. However, the multiple bouncing of the diffracted fields from the walls and the ground is ignored in the GO+UTD code due their negligible effects compared to strong reflections. Therefore, a little discrepancy among the results appears there, as expected.

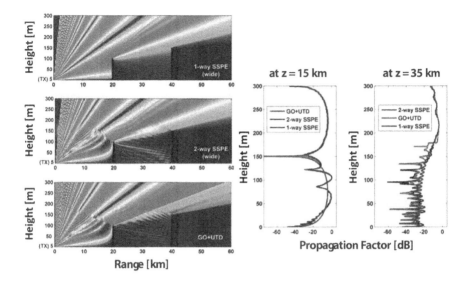

Figure 8.6   PF vs. range/height over flat Earth with double knife edges (100 m height-wall at 20 km range and 150 m height-wall at 40 km range) with horizontal polarization: (Left: from top to bottom) 1W-SSPE, 2W-SSPE, GO+UTD, (right) PF vs. height at two specified ranges; $f = 3$ GHz, $\Delta x = 0.29$ m, $\Delta z = 200$ m.

The scenario presented in Fig. 8.6 involves two finite-height walls along the propagation path (100 m height wall at 20 km range and 150 m height wall at 40 km range) illuminated by a line source at 5 m (i.e., multiple-wedge problem). The walls and the ground satisfy DBC. The three field maps (from top to bottom) belong to 1W-SSPE, 2W-SSPE, and GO+UTD results. The boundaries of both incident (i.e., LOS) and reflected fields are observed in both 2W-SSPE and GO+UTD maps. The artificial effects around these boundaries are also clearly observed in the GO+UTD map (this is not observed in the SSPE maps). The two plots on the right side belong to PF vs. height at two different ranges showing the comparison among the three models. As observed from 2W-SSPE and GO+UTD results, a good agreement is obtained in the first region (at 15 km, i.e., in the interference region), but a slight discrepancy is observed in the shadow region (at 35 km, i.e., in the shadow/diffraction region). This discrepancy can be eliminated by taking into account the slope diffraction coefficient in the GO+UTD model, whose accuracy decreases in deep-shadow regions. Furthermore, this discrepancy might be due to ignoring some of the less contributing components (such as double diffractions), and due to the limitation of the SSPE within the paraxial regions, therefore further investigation is required in order to speculate about this discrepancy.

8.5 Two Way Propagation Modeling in Waveguides

Parabolic equation models, since their first introduction [18] have long been used in propagation through complex environments. They are based on the assumption of slow longitudinal variations and are derived from the wave equation that yields one-way, forward marching solutions. A reference list is never complete because of the size of the literature on PE, but [3–5] might be a good start. Some of the recent studies and references therein may also be useful [31, 34, 48]. Finite-difference, finite-element, and FFT-based algorithms have been used in 2D (range–height) space, and the effects of ground losses/irregularities and environmental/atmospheric inhomogeneities have been successfully modeled.

3D PE algorithms have also been introduced and applied in propagation modeling through open and closed environments [98, 100–102]. Moreover, two-way PE models have been developed for the propagation problems where backward scattering is significant [27, 67, 68, 97].

The 2W-SSPE and 2W-FEMPE algorithms, which were developed, tested, and calibrated for the groundwave propagation in the presence of knife edge obstacles [27] as well as arbitrary-shaped, irregular terrain profiles, are modified and used in modeling propagation inside rectangular waveguides having obstacles at various ranges. The novel 2W-SSPE and 2W-FEMPE algorithms are compared to each other as well as calibrated against analytical reference data.

Consider 3D propagation inside a rectangular waveguide determined by its width (a), height (b), and located longitudinally along the z-direction. The initial source distribution inside a rectangular waveguide can be represented in terms of modal summation (see Chapter 7 for details).

Terminating the waveguide completely at any range along z-direction yields total reflections inside and two-way interference can be modeled by using image method in order to test numerical PE models. This procedure works by removing the obstacle at range z_t and putting an image source at range $2z_t$.

8.6 Three-Dimensional Split-Step- and Finite-Element-Based Parabolic Equation Models

The 3D PWE for rectangular waveguides homogeneously filled with air uses the standard PE given as

$$\left(\frac{\partial^2}{\partial x^2} + \frac{\partial^2}{\partial y^2} + 2ik_0 \frac{\partial}{\partial z} \right) u(x, y, z) = 0 \qquad (8.1)$$

where x and y stand for the transverse coordinates, z is the longitudinal coordinate, and $u(x, y, z) = \exp(-ik_0 z)\psi(x, y, z)$ is the reduced function since the direction of wave propagation is predominantly along the z-axis.

The idea of SSPE model is to solve (8.1) in the transverse k_x and k_y spectral-domains. In the numerical computations, this is achieved by using FFT. The BCs are imposed manually. The choice of discrete step size (Δx, Δy, and Δz), the injection of transmitting antenna, implementation of boundary losses as well as irregularities are well known and can be found in many sources (e.g., see [3, 5, 31, 97]).

The idea of FEMPE model is to divide the 2D transverse-domain into subdomains first. Then, the initial field profile, generated from a Gaussian antenna pattern, is injected. The approximated field values at discrete nodes are then propagated longitudinally by using the Crank–Nicolson approach, which is based on improved Euler method with DBC (NBC) at each range for horizontal (vertical) polarization, respectively [31, 97].

The modified PE models can handle two-way propagation effects. The modification is based on the generation of backward-reflected waves when an obstacle is met. In other words, the wave is marched out in the forward direction until it hits an obstacle and is separated into two components; forward propagating and backward propagating, by imposing the appropriate BCs.

The forward wave continues after setting the field profile of obstacle region zero. Then, the backward field is obtained by providing the BCs for the obstacle. Since the total tangential field must be zero for PEC case, the backward function can be found as $u_b(x, y, z) = \exp(ik_0 z)\psi_b(x, y, z)$. Finally, the total field is obtained by superposing both fields at each range. For multiple obstacles and when resonances occur, multi-forward and multi-backward generation is applied. The number of reflections in both directions is then determined from the stated accuracy.

8.7 Tests and Calibration

The first test belongs to a square waveguide (4 m $\times$ 4 m) terminated at 200 m range. The wave function inside the PEC waveguide with TM waves is represented in terms of modal superposition. The frequency is chosen as 3 GHz (well above the cutoff frequency). The transverse discretization values are $\Delta x = 0.8\lambda$ and $\Delta y = 0.8\lambda$, and the longitudinal step size is $\Delta z = 10\lambda$. A 2D Gaussian tilted antenna with an elevation of $\theta_{(elvx, elvy)} = (1°, 2°)$ and beamwidth of $\theta_{bw} = 4.34°$ is used. The source position is $x_s = 2$ m, $y_s = 2$ m, and $z_s = 0$. The number of modes in the transverse-domain are $N_x = 21$ and $N_y = 22$ along x and y directions (for a relative maximum error of 1e-7). Note that the relative maximum error is defined as the maximum difference in absolute values of the Gaussian field and the modal summation divided by the maximum absolute value of the Gaussian field. Perfect agreement is obtained among three models. Figure 8.7 shows 2D cross-sectional field distributions for this case. Very good agreements are clearly obtained. Fields vs. range at point $x_r = 2$ m, $y_r = 1$ m for tilted and untilted sources are also plotted. As observed, there is a very good agreement among the modal solution, SSPE, and FEMPE solutions. Note that SSPE, compared to FEMPE, gives better results because of the match of the DST and sinusoidal functions of modal summation.

Using the same parameters above, the second test belongs to one 4 m $\times$ 2 m obstacle at 100 m range. The region up to 100 m exhibits forward-propagating waves interfering with backward waves. Figure 8.8 shows the cross-sectional field distri-

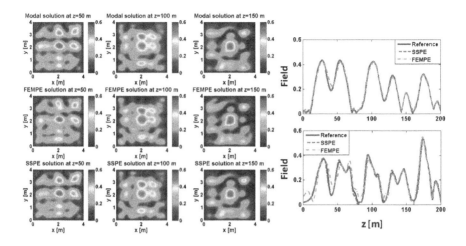

Figure 8.7 (Left) 2D cross-sectional field distributions inside PEC rectangular waveguide terminated at 200 m range. Rows: modal summation, FEMPE, and SSPE simulations. Columns: 50 m, 100 m, and 150 m ranges, respectively. (Right) 1D longitudinal field distributions at $x_r = 2$ m, $y_r = 1$ m: (top) untilted, (bottom) tilted: $\theta_{(elvx, elvy)} = (1°, 2°)$.

butions of two methods at three different ranges. The rows correspond to 1W-SSPE, 2W-SSPE, and 2W-FEMPE; the columns belong to ranges of 80 m, 90 m, and 150 m, respectively. The difference between one-way and two-way models is clearly seen, and both numerical techniques give similar results. Note that the field distributions at 150 m range are similar because only forward waves contribute in this region. The transmitted power is obtained as 48.4% and 48.3% for SSPE and FEMPE methods, respectively. Fields vs. range at point $x_r = 2$ m, $y_r = 1$ m for one-way and two-way FEMPE and SSPE results are also given. As observed, the difference between one-way and two-way propagators is significant before the obstacle. Beyond that field distributions are exactly the same because only forward waves contribute in this region.

Finally, Figs. 8.9 and 8.10 belong to arbitrary terrain profiles which do not have any analytical exact or approximate solution that might serve as a reference, where only FEMPE model can be compared with the SSPE model for horizontal polarization. Figure 8.9 shows the comparison of narrow- and wide-angle FEMPE results through homogeneous standard atmosphere. Two Gaussian antennas having elevations of 25° and 5° downward are at 150 m and 200 m heights, respectively. The beamwidth is the same for both as $\theta_{bw} = 1°$, and the frequency is 600 MHz. The transverse discretization is $\Delta x = 0.42$ m and the longitudinal step size is $\Delta z = 1$ m. As observed, narrow-angle FEMPE cannot accurately account for the reflections which change the true field map considerably.

3D field maps vs. range/height of another scenario through inhomogeneous atmosphere produced with the 2W-FEMPE and 2W-SSPE models are given in Fig. 8.10. The source is at 400 m height with a beamwidth of 0.5° and an elevation of 0.5°

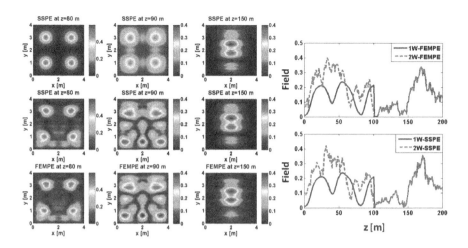

Figure 8.8 (Left) 2D cross-sectional field distributions of untilted source inside PEC rectangular waveguide with one 4 m × 2 m obstacle at 100 m range. Rows: 1W-SSPE, 2W-SSPE, and 2W-FEMPE simulations. Columns: 80 m, 90 m, and 150 m ranges, respectively. (Right) 1D longitudinal field distributions at $x_r = 2$ m, $y_r = 1$ m: (top) FEMPE, (bottom) SSPE.

downward at 300 MHz. The inhomogeneous atmosphere has a refractivity slope of -200 M units/km between 300 m and 400 m (ducting case). The transverse discretization is $\Delta x = 1$ m and the longitudinal step size is $\Delta z = 200$ m. Note that Fig. 8.10 needs further comparisons against another full wave (e.g., the MoM-based [11, 42, 103] and/or the time-domain propagation [13, 14, 104, 105]) models in order to complete the VV&C tests. This is because both SSPE and FEMPE approaches start with the same PE model and equally inherit all modeling incapabilities (e.g., the paraxial approximation). Yet, analytical exact solutions have not appeared for the irregular terrain models. The asymptotic GO+UTD model is not applicable (in its current form) to problems containing irregular terrains.

The FEMPE model needs much smaller range steps than the SSPE model, therefore necessitates longer computation times when long-range propagation is of interest. According to the calculations, the computation time of the FEMPE is at least 14 times longer as compared to that of SSPE. As a result, the number of nodes used in the transverse and range coordinates should be optimized with respect to the following parameters: the operating frequency, the irregular terrain structure, inhomogeneous atmosphere, and the selection of the initial Gaussian antenna pattern specified by its height, beamwidth, and tilt. Moreover, the line-source excitation is a serious problem in the FEMPE model, but this might be overcome by constructing the initial field profile at a few wavelengths away from the line source by using some other models such as the SSPE or GO, or using smooth, such as Gaussian type, source patterns. On the other hand, FEMPE model handles all kinds of BC at the surface much more easily when compared with the SSPE model.

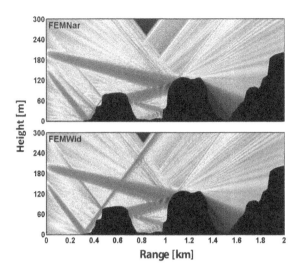

Figure 8.9 PF vs. range/height over arbitrary terrains through homogeneous standard atmosphere.

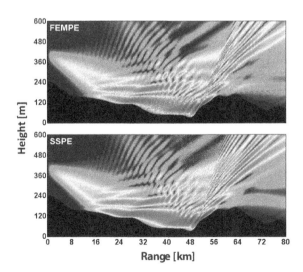

Figure 8.10 PF vs. range/height over arbitrary terrains through inhomogeneous atmosphere.

CHAPTER 9

PETOOL VIRTUAL PROPAGATION PACKAGE

9.1 Introduction

Radio wave propagation over the Earth's surface and in an inhomogeneous atmosphere is affected by several scattering phenomena, such as reflection, refraction, and diffraction. Understanding the effects of varying conditions on radio wave propagation is, therefore, essential for designing reliable radar and communication systems. Especially, tropospheric waves may play a dominant role in communications because they can propagate over the horizon and increase the coverage area, and hence, they may disrupt the communication links due to the interference that is not normally there. Such waves are propagated by bending or refraction due to the abrupt change in the refractive index in the troposphere and cause so-called *anomalous* propagation. If the refractive gradient exceeds some certain limits, the radio waves may be trapped in a *duct* and guided over distances far greater than the normal range. In addition to tropospheric effects, irregular terrain surfaces have considerable influence on radio wave propagation because they reflect and diffract the EM waves in a complex way. Hence, the design of an effective radar or communication system can be achieved by using a model that can properly incorporate the refractivity and terrain factors.

Radio Wave Propagation and Parabolic Equation Modeling, First Edition. By Gökhan Apaydin, Levent Sevgi
© 2017 by the Institute of Electrical and Electronic Engineers, Inc. Published 2017 by John Wiley & Sons, Inc.

The rigorous analytical and numerical modeling of radio wave propagation in such environments is a challenging task and has attracted the attention of researchers for many decades. The difficulty stems from the vast variability of the properties of the medium and also the surfaces and obstacles that re-direct the propagating energy, making the radio propagation somewhat unpredictable. Initially, analytical techniques (such as ray tracing methods, diffraction methods, and waveguide mode theory) have been employed to predict the radio propagation [39, 78, 80, 91, 99, 106]. However, they require the geometry to be represented as a member of a set of some canonical geometries and suffer from the presence of the vertically varying refractivity profile in the troposphere. With the advances in computers, some numerical techniques have been devised to easily handle the above-mentioned difficulties. PE model has been widely used in propagation modeling to predict the wave behavior between a transmitter and a receiver over the 2D Earth's surface, because of its high capability in modeling both horizontally and vertically varying atmospheric refraction (especially ducting) effects. The standard PE is derived from Helmholtz's equation in such a way that the rapidly varying phase term is discarded to obtain a reduced function having slow variation in range for propagating angles close to the paraxial direction. Helmholtz's equation is approximated by two differential equations, corresponding to forward- and backward-propagating waves, each of which is in the form of a parabolic PDE. The standard PE method takes only the forward part into account, namely, it is a one-way, forward scatter model, valid in the paraxial region. Although the initial introduction of the PE method has been credited to [18], its wide-spread usage has become possible after the development of Fourier split-step algorithm by [107]. The SSPE is, in general, an initial-value problem starting from a reference range (typically from an antenna), and marching out in range by obtaining the field along the vertical direction at each range step, through the use of step-by-step Fourier transformations [3, 5, 10, 60, 62, 108]. Apart from the Fourier split-step algorithm, the solution of the PE has also been achieved by the finite-difference-based [56, 109, 110] and finite element (FE)-based [22, 27, 31, 44, 48] algorithms. The Fourier split-step algorithm is more robust since it provides the use of larger range increments and a faster solution for long-range propagation scenarios. Apart from these studies, there exist several computer software programs, most of which have been developed for military purposes, for predicting radar coverage under the effect of environmental factors that influence refractivity. These are IREPS (Integrated Refraction Effects Prediction System), EREPS (Engineer's Refractive Effects Prediction System), TESS (Tactical Electronic Support System), AREPS (Advanced Refractive Effects Prediction System), TEMPER (Tropospheric Electromagnetic Parabolic Equation Routine), and TPEM (Terrain Parabolic Equation Model). These programs implement the ray optics or one-way PE techniques, or a hybrid model combining these methods.

In this chapter, a novel software tool (PETOOL), which is developed in MATLAB with graphical user interface (GUI), is introduced for the analysis and visualization of radio wave propagation through the homogeneous and inhomogeneous atmosphere, by incorporating variable terrain effects with the aid of the two-way split-step algorithm employing wide-angle propagator. The reason to develop another PE-based program for radio wave propagation is twofold: The first is that, PETOOL is a free open-source program, and has been designed with a user-friendly GUI, to serve as a research/educational tool for propagation engineers/instructors to investigate the phenomenon in an illustrative manner, and/or to achieve the analysis/design/planning of reliable communication systems. It displays the propagation factor/loss on a range/height scale and enables the user to easily visualize, enter, and save all related input/output parameters. More importantly, the user can easily create arbitrary terrain and refractivity profiles. The second reason is that PETOOL is indeed the first software package implementing both 1W-SSPE and 2W-SSPE algorithms, the latter of which incorporates the multipath effects into the PE solution of the radio wave propagation through a recursive forward–backward algorithm. The standard one-way PE method, in spite of its wide-spread usage, suffers from two major drawbacks.

- The PE method handles only the forward-propagating waves and neglects the backscattered waves. The forward waves provide almost accurate results for typical long-range propagation problems, only if there does not exist obstacles that redirect the incoming wave in the form of reflections and diffractions. However, the accurate estimation of the multipath effects, occurring during propagation over terrain, requires the correct treatment of backward waves as well. Moreover, the PE method takes the diffraction effects into account within the paraxial approximation, degrading the accuracy of the approach in deep-shadow regions where the diffracted fields dominate. There are a number of studies in the literature to overcome such difficulties [63, 64, 67, 111–115]. A two-way PE algorithms have also been proposed (see Chapter 8 and [68, 116]).

- The second drawback is that the standard PE is a narrow-angle approximation, which consequently restricts the accuracy to propagation angles up to $10° - 15°$ from the paraxial direction. A typical long-range propagation encounters propagation angles that are usually less than a few degrees, whereas the short-range propagation problems, as well as the problems involving multiple reflections and diffractions because of hills and valleys with steep slopes, can only be solved by a PE model that is effective for larger propagation angles. To handle propagation angles up to $40° - 45°$, wide-angle propagators have been introduced [25, 26, 65, 117–119]. PETOOL implements the two-way algorithm proposed in [68], with the exception that narrow-angle propagators are replaced by wide-angle propagators to handle larger propagation angles.

9.2 PETOOL Software

PETOOL software package [96] has been developed in MATLAB with a user-friendly GUI for the analysis and visualization of radio wave propagation. The GUI has been designed so as to meet the following goals:

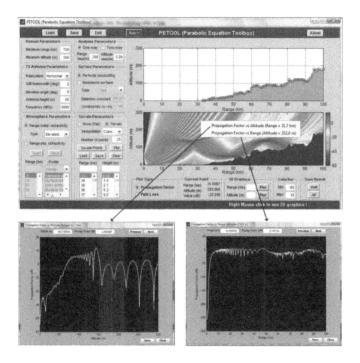

Figure 9.1 PETOOL main window and 2D graphics windows obtained by right-clicking the mouse on the 3D map.

- The user should be able to easily visualize, load, and save the propagation factor/loss on a range/height scale in radio wave propagation problems over variable terrain and through the homogeneous and inhomogeneous atmosphere.

- The user should be able to define her/his own input parameters, and to load/save them if desired. The user should be warned if s/he enters inappropriate input values.

- The user should be able to easily define an arbitrarily shaped terrain profile by just locating a number of points on the graph through left-clicking the mouse. The user should be able to load/save the terrain parameters from/in a user-defined file.

- The user should be able to easily specify range dependent or range independent refractivity profiles by just selecting from a list of various types of atmosphere profiles.

The main m-file to run the program is *petool.m*. The main window of the program is depicted in Fig. 9.1. The window is basically divided into four panels. The left and right panels are located on a large gray background, whereas the top and bottom panels are located on thin blue backgrounds. These panels are defined in detail below.

The *top blue panel* is reserved for five operational push buttons (load, save, exit, run, about). The load and save buttons are used for all input parameters of the simulation. Once clicked, a modal dialog box is opened to select or specify a file the user wants to create or save. While exiting PETOOL, the user is also warned whether or not s/he wants to save the parameters.

The *bottom blue panel* is used to show warning text messages whenever needed, especially in the case of inappropriate input entrance.

On the *left panel*, there are six subpanels (domain, analysis, antenna, surface, atmosphere, terrain), where the input parameters are defined by the user. The input parameters are summarized in Table 9.1.

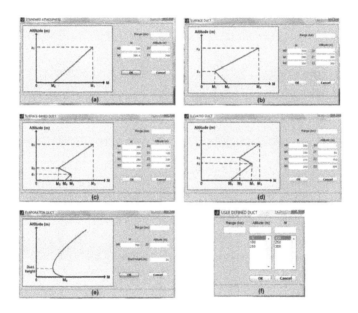

Figure 9.2   Windows for specifying atmosphere types: (a) standard atmosphere, (b) surface duct, (c) surface-based duct, (d) elevated duct, (e) evaporation duct, (f) user-defined duct.

- In specifying a refractivity profile, the user selects an atmosphere type from a menu list, and a new window is opened accordingly, enabling the user to enter the modified refractivity (M) values for the specified atmosphere type. The available atmosphere profiles are the standard atmosphere, surface duct, surface-based duct, elevated duct, evaporation duct, and user-defined duct, whose windows are illustrated in Fig. 9.2. If the range dependent refractivity profile is to be defined, the above-mentioned selection is performed for each range value. The profiles lying between two consecutive range values are computed automatically through linear interpolation. It is useful to note that the user can load/save the atmosphere parameters separately. In addition, the user can easily modify or

Table 9.1   Input parameters of PETOOL.

Domain Parameters:

- Maximum range [km]   • Maximum height [m]

Antenna Parameters:

- Polarization (horizontal or vertical)   • 3 dB beamwidth [degree]
- Antenna height [m]   • Elevation angle [degree]
- Frequency [MHz] (must be >30 MHz)

Analysis Parameters:

- Range step [m]: horizontal step size   • Height step [m]: vertical step size
- One way or two way

(which algorithm is to be performed)

Surface Parameters:

- PEC or impedance surface   If the user-defined ground is chosen:

If impedance surface is chosen: • Dielectric constant

Type: Sea, fresh water, wet ground,   • Conductivity [S/m]

medium dry ground, very dry ground,

user-defined ground.

Atmosphere Parameters:

- Range independent or

range dependent refractivity

If range independent refractivity is chosen: If range dependent refractivity is chosen:

Type: standard atmosphere, surface duct,   Refractivity type for each range value

surface-based duct, elevated duct, evaporation   defined in list boxes

duct, and user-defined duct.

Terrain Parameters:

- None (flat surface) or terrain

If terrain is chosen:

- Interpolation type (none, linear, cubic spline)   • Number of points: number of points
- Range and height values for terrain points   to be placed on top graphics to define

defined in list boxes   the terrain profile

delete the parameters in the profile and range lists, by means of a special dialog box that is opened when the user clicks an item from the list.

- In specifying a terrain profile, the user has three options: (i) s/he can locate a number of points on the top graphics of the right plane by clicking *Locate Points* button, (ii) s/he can define the terrain points manually by entering the values into the range–height list boxes, or (iii) s/he can load a user-generated text file including the terrain parameters. In all cases, the user can save/load/clear/plot the terrain profile. If the user prefers to create her/his own terrain by locating points on the graphics, the values of the selected points are automatically placed into the range and height list boxes. Hence, it is possible to store the graphically generated terrain profiles in files, as well as to modify or delete the parameters in the list boxes. Once the terrain points are specified by using one of the above-mentioned ways, the overall terrain profile is created by performing a kind of interpolation (linear or cubic spline) between two consecutive terrain points along the range. If the interpolation method is chosen to be *none*, the terrain profile appears as a collection of knife edges. After defining the terrain profile, the user must click on the *run* button to see the analysis results of the new geometry. It is useful to note that the program does not allow the terrain to extend below zero level.

On the *right panel*, there are two graphics (top graphics where the terrain is specified, and the bottom graphics showing the colored 3D map of the propagation factor/loss), together with five sub-panels (plot type, current point, 2D graphics, colorbar, save result) related to the visualization or storage of output parameters. After the user clicks on the *run* button, the code performs the 1W-SSPE or 2W-SSPE algorithm, and then, plots the 3D PF map on the bottom graphics. Although the default is PF, the user can switch to PL map by clicking the appropriate button in the *plot type* panel. Whenever the user moves the mouse over the bottom graphics, the values (range, height, PF/PL) automatically appear in the *current point* panel. The user can plot the 2D graphics (PF/PL vs. range for fixed height, or PF/PL vs. height for fixed range) either by entering the values into the boxes in *2D graphics* panel, or by right-clicking the mouse on the desired point of the 3D map (see Fig. 9.1). The *colorbar* panel is used to adjust the colorbar scale of the 3D map for better visualization. Finally, the *save result* panel is used to store the PF/PL maps in the form of a MATLAB file (.mat) or a picture file (.tif). We also note that 2D graphics in Fig. 9.1 can be saved in text files by using the *save* button.

9.3 Characteristic Examples

It might be difficult to produce numerical reference data for the VV&C tests. A convenient way of testing narrow- and wide-angle SSPE models is to tilt up or down the pattern up to $40° - 45°$. In the first scenario, five Gaussian antennas located at the same place (1000 m height) with 3 dB beamwidth of $0.5°$ each, and $10° - 20° - 30° - 40° - 50°$ tilt angles illuminate the PEC ground in a standard atmosphere, assuming that the frequency is 1 GHz, the polarization is horizontal, the range and height increments are $\Delta z = 10$ m and $\Delta x = 0.13$ m, respectively. Three-dimensional PF maps, corresponding to narrow- and wide-angle SSPE are illustrated in Fig. 9.3. Although there is an almost exact match between the desired specular reflection points and those found by the wide-angle SSPE, the results of the narrow-angle SSPE start to deteriorate as the tilt angle increases.

Next, the performance of the 1W-SSPE is demonstrated against AREPS, which simulates only the one-way propagation with narrow-angle SSPE. As shown in Fig. 9.4,

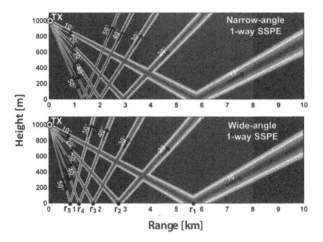

Figure 9.3 PF maps for five Gaussian antennas with different tilt angles illuminating PEC ground: (top) narrow-angle, (bottom) wide-angle.

an untilted Gaussian antenna at 50 m height with 3 dB beamwidth of $3°$ illuminates a variable terrain inside a surface duct, which is modeled by the modified refractivity profile M. The frequency is 3 GHz. The discretization parameters are $\Delta z = 200$ m and $\Delta x = 0.29$ m. As illustrated by the PF maps, a good agreement is obtained between PETOOL and AREPS in the narrow-angle case. The good agreement among the results illustrates the success and completeness of the VV&C process for the 1W-SSPE.

In the following two scenarios, the calibration is done against the GO+UTD results, assuming that the frequency is 3 GHz. It is useful to note that the calibration via GO+UTD is feasible only for sufficiently high frequencies such that the ray-optics interpretation is valid.

The first scenario involves two walls, one of which is finite (also known as a knife edge) and the other one is infinite in height. In Fig. 9.5, the finite-height (50 m) wall is located at 50 km; the infinite-height wall is at 60 km; the line source is at 250 m, and the polarization is vertical. The discretization parameters are $\Delta z = 200$ m and $\Delta x = 0.29$ m. This is one of the classical structures in the field of diffraction theory, and its approximate solution can be computed by using ray-optic techniques, combined with special diffraction methods. In the implementation of the GO method, the reflected waves from the ground and the wall can simply be calculated by employing the principles of image theory that replaces the original problem with the equivalent problem represented by image sources with respect to the BCs that must be satisfied on the boundaries depending on the polarization. The total field is obtained by the sum of the direct ray, reflected rays emanating from image sources, and the diffracted rays from the tip of the wall, by also checking the LOS conditions between the source(s) and the observation point. As shown, multiple reflections (almost resonance behavior) occur especially in the region between the walls. In

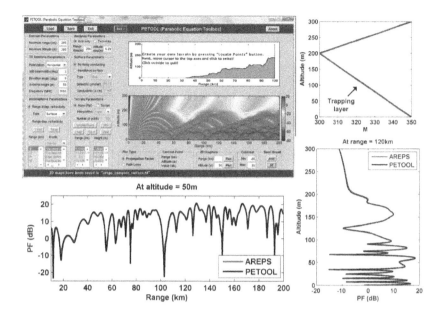

Figure 9.4 (Upper-left) Main window with PF map, (upper-right) modified refractivity profile for surface duct, (lower-left) PF vs. range at 50 m height, (lower-right) PF vs. height at 120 km range.

comparing the 2W-SSPE with the GO+UTD approach, the contribution of the waves hitting the walls up to three times is superposed. To achieve fair comparisons up to third degree of reflections, the GO+UTD code accounts for 35 types of rays bouncing from the walls and the ground. In addition to reflected waves, the diffracted waves from the finite-height wall are also computed. However, the multiple bouncing of the diffracted fields from the walls and the ground is ignored due to their negligible effects compared to strong reflections. The good agreements among the results illustrate the success of the 2W-SSPE with respect to GO+UTD approach.

The next scenario in Fig. 9.6 illustrates the comparison between 1W-SSPE and 2W-SSPE models with narrow- and wide-angle propagators over a variable terrain and in an elevated duct environment. A Gaussian antenna with $2°$ tilt angle is located at 50 m height. The frequency is 3 GHz, and the polarization is horizontal. The discretization parameters are $\Delta z = 200$ m and $\Delta x = 0.29$ m. Referring to the descriptions in Fig. 9.2, the elevated duct is defined as follows: $M_0 = 300$, $M_1 = 330$, $M_2 = 310$, $M_3 = 350$, $x_1 = 100$ m, $x_2 = 150$ m, $x_2 = 300$ m. These examples help us to visualize the importance of the correct representation of large propagation angles, as well as the backward-propagating waves.

As emphasized in the introduction part, PETOOL has been designed for research/educational purposes. Hence, propagation engineers/instructors can simulate different propagation scenarios to investigate the radio wave propagation and/or to design reliable communication links. In order to demonstrate what engineer/instructors can

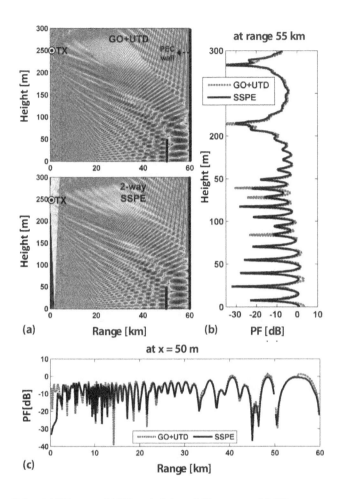

Figure 9.5   (a) PF maps, (b) PF vs. height at 55 km range, (c) PF vs. range at 50 m height.

do with PETOOL, we have shown various scenarios at 3 GHz in Figs. 9.7 and 9.8. The discretization parameters are $\Delta z = 200$ m and $\Delta x = 0.29$ m. In Fig. 9.7, a Gaussian antenna, which is located at 10 m height, radiates into the surface duct and surface-based duct environments. Referring to the descriptions in Fig. 9.2, the surface duct is defined as follows: $M_0 = 350$, $M_1 = 300$, $M_2 = 350$, $x_1 = 200$ m, $x_2 = 300$ m. The surface-based duct is defined as follows: $M_0 = 340$, $M_1 = 356$, $M_2 = 340$, $M_3 = 358$, $x_1 = 135$ m, $x_2 = 150$ m, $x_2 = 300$ m. Polarization is horizontal, and the ground is PEC. Note that the computation times are dependent on the number of iterations in the 2W-SSPE, which in turn depend on the amount of the wave interactions (multiple reflections) between the hills.

In Fig. 9.8, different ground parameters are simulated assuming that a Gaussian antenna is located at 5 m height and radiates into a standard atmosphere. On the

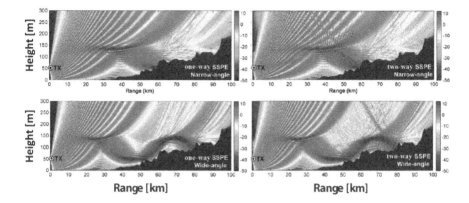

Figure 9.6   PF maps for a variable terrain in an elevated-duct environment: (left) 1W-SSPE, (right) 2W-SSPE, (top) narrow-angle, (bottom) wide-angle.

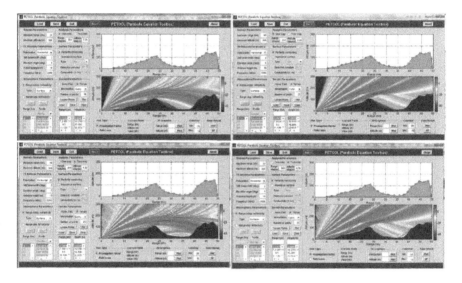

Figure 9.7   PF maps for a PEC terrain illuminated by a Gaussian antenna at 10 m assuming different refractivity profiles: (left) surface duct, (right) surface-based duct, (top) 1W-SSPE, (bottom) 2W-SSPE.

left side, the surface is PEC, on the right side the surface is *very dry ground* with dielectric constant $\varepsilon_r = 3$ and conductivity $\sigma_g = 1e - 4$ Siemens/m. The discretization parameters are $\Delta z = 100$ m and $\Delta x = 10$ m. Polarization is vertical and the frequency is 100 MHz.

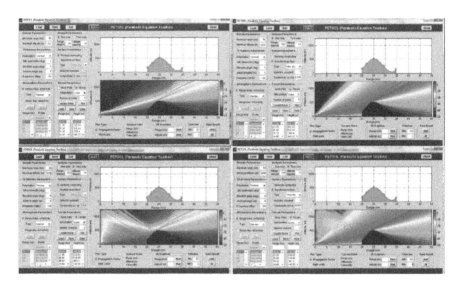

Figure 9.8   PF maps for a variable terrain illuminated by a Gaussian antenna at 5 m assuming different ground surface parameters in standard atmosphere: (left) PEC surface, (right) very dry ground, (top) 1W-SSPE, (bottom) 2W-SSPE.

CHAPTER 10

FEMIX VIRTUAL PROPAGATION PACKAGE

10.1 Introduction

The ITU regulates EM groundwave propagation prediction [120, 121]. Recommendation P-368-7 [121] fully covers both *homogeneous* and *mixed path* groundwave propagation problems. It gives a set of curves of predicted field strength as a function of distance for vertically polarized EM waves in the medium-frequency (MF) and high-frequency (HF) bands for a variety of ground conductivity (σ_g) and relative permittivity (ε_r) values [9]. These curves are valid for a certain range of antenna heights for a homogeneous atmosphere up to heights $h = 1.2\sigma_g^{1/2}\lambda^{3/2}$ if $\varepsilon_r \ll 60\sigma_g\lambda$. They are also valid for non-flat longitudinal paths with obstacles not higher than a wavelength and not closer than 8–10 km to either the transmitter or the receiver [9]. Systems such as Digital Radio Mondiale (DRM) and/or high-frequency surface-wave radars (HFSWRs) [122, 123] necessitate revisiting groundwave propagation modeling and the mixed path problem. Among available numerical packages, GRWAVE and DRMix [15] are worthwhile mentioning. These packages use Norton's [50] and Wait's [51] models, as well as Millington's curve fitting approach [52]. All these models can handle only smooth spherical Earth problems. [3, 4, 9, 16, 124] should be referred for a summary of most of these early analytical studies. Note that surface-wave propagation along inhomogeneous paths can be significantly affected by the surface losses between 30 kHz and 30 MHz (LF/MF/HF).

Surface-wave PL vs. range have been predicted via EM wave propagation modeling along homogeneous/inhomogeneous propagation paths over irregular terrain. An *irregular terrain* means a change of land irregularities above Earth's surface. A *homogeneous propagation path* means the use of the same impedance BC at the Earth's surface (lake, sea, dry land, wet land, etc.). An *inhomogeneous propagation path* means having different impedance BCs on the surface along the range (e.g., sea–land–sea, land–lake–land, transitions).

The groundwave propagation problem over a non-smooth spherical Earth with an impedance BC and a radially dependent atmosphere, excited by a vertical electric dipole located near the Earth's surface, requires the solution of a 3D EM wave equation in spherical coordinates. Unfortunately, a 3D solution in terms of known numerically computable mathematical functions has not appeared yet. All available analytical solutions are in 2D transverse–longitudinal space, assume azimuthal asymmetry, and are based on either ray or mode summation methods [124].

A novel FEMPE-based software tool (FEMIX) is introduced for the analysis and visualization of surface-wave propagation over the irregular Earth's surface through a homogeneous and an inhomogeneous atmosphere. A typical scenario is pictured in Fig. 10.1. Here, z and x represent the range and height coordinates, Δz and Δx are the range and height step sizes, respectively. The 2D environment is open along z (i.e., $-\infty < z < +\infty$) and semi-open along x (i.e., $0 < x < +\infty$). The propagation problem is postulated via 2D Helmholtz's equation, together with the appropriate termination conditions (i.e., an impedance BC at $x = 0$ and radiation condition for $z \to \pm\infty$ and $x \to +\infty$).

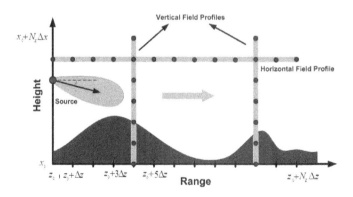

Figure 10.1 The 2D radio wave propagation scenario over irregular terrain. The next vertical field is obtained from the previous field via step-by-step iterative marching procedure.

10.2 Analytical Surface-Wave Model

The analytical surface-wave model over inhomogeneous smooth spherical Earth's surface paths, recommended by the ITU, uses a ray-optical Norton's surface-wave [50] and Wait's surface-guided mode [51] models for the LOS and the shadow regions, respectively; and the Millington curve fitting method [52].

10.2.1 Path Loss

The PL is the ratio between the power radiated by the transmitting antenna (P_t) and the power available at a point in space (P_r) [121]

$$PL(d) = 10 \log \frac{P_r(z_s + d)}{P_t(z_s)} \qquad (10.1)$$

which is calculated at a given distance d as

$$PL(d) = 142 + 20 \log(f_{MHz}) + 20 \log(E_{\mu V/m}) \qquad (10.2)$$

by taking the received power $P_r = E_r^2 \lambda^2 / (Z_0 4\pi)$ where E_r is the received field strength at distance $z = z_s + d$, λ is the wavelength, $Z_0 = 120\pi$ is the free-space wave impedance, and $P_t = 1$ kW at the initial range.

10.2.2 Norton's Model

Under the standard atmosphere and spherical-smooth Earth assumptions, Norton's model uses the ray-optical asymptotic approximation from a wavenumber spectral integral representation [50]. Considering the close placement of the source (h_t) and the receiver (h_r) to the ground (at short and medium ranges, where flat Earth assumption can be made), the normal and tangential electric field components of the surface-wave are

$$E_n = (1 - R_v)E_0(1 - u^2 + u^4 \cos^2 \psi)F(\kappa) \qquad (10.3)$$

$$E_t = (1 - R_v)E_0(1 - u^2 \cos^2 \psi)^{1/2}(1 + 0.5 \sin^2 \psi)F(\kappa) \qquad (10.4)$$

where $E_0 = ik_0 Z_0 M_0 \exp(ik_0 d)/(4\pi d)$, $\psi = \tan^{-1}((h_t+h_r)/d)$, $M_0 = 5\lambda/2\pi$ Am is the dipole moment for a short electric dipole [121], $u^2 = 1/\varepsilon_g$, $\varepsilon_g = \varepsilon + i60\sigma_g\lambda$ is the relative dielectric constant of the surface, and

$$R_v = \frac{\varepsilon_g \sin \psi - \sqrt{\varepsilon_g - \cos^2 \psi}}{\varepsilon_g \sin \psi + \sqrt{\varepsilon_g - \cos^2 \psi}} \qquad (10.5)$$

is the reflection coefficient for vertical polarization. The attenuation function is

$$F(\kappa) = 1 - i\sqrt{\kappa\pi} \exp(-\kappa)\mathrm{erfc}(i\sqrt{\kappa}) \qquad (10.6)$$

$$\text{erfc}(i\sqrt{\kappa}) = \frac{2}{\sqrt{\pi}} \int_{i\sqrt{\kappa}}^{\infty} \exp(-t^2)dt \qquad (10.7)$$

$$\kappa = -\frac{ik_0 d}{2} u^2 (1 - u^2 \cos^2 \psi) \left(1 + \frac{\sin \psi}{u\sqrt{1 - u^2 \cos^2 \psi}} \right) \qquad (10.8)$$

where $\text{erfc}(i\sqrt{\kappa})$ is the complementary error function. Norton's model is good within the LOS region.

10.2.3 Wait's Model

Wait's model uses the spectral integral as a series of normal modes propagating along a spherical Earth's surface [51, 121]. Under the flattened-Earth assumption (i.e., when the Earth's curvature is modeled with the effective radius), the normal component of the electric field is

$$E_n = E_0 F(x, x'; z) \qquad (10.9)$$

with the attenuation function

$$F(x, x'; z) = \frac{\pi z}{2} \sum_{s=1}^{\infty} \frac{\exp(i\beta_s z)}{\beta_s - q^2} \frac{W(\beta_s - x)}{W(\beta_s)} \frac{W(\beta_s - x')}{W(\beta_s)} \qquad (10.10)$$

where x' and x are the source and observation heights, $q = imn_0 Z_s / Z_0$, $m = (k_0 a_e / 2)^{1/3}$, n_0 is the refractive index at the Earth's surface, a_e is the Earth's radius, and the surface impedance is

$$Z_s = Z_0 \left(\frac{1}{\varepsilon - i60\sigma_g \lambda} \right)^{1/2} \left(1 + \frac{1}{\varepsilon - i60\sigma_g \lambda} \right)^{1/2}. \qquad (10.11)$$

For the standard atmosphere including the Earth's curvature, the transverse mode functions are the solutions of Airy's equation [3],

$$W(\beta) = \sqrt{\pi} \left(Bi(\beta) - iAi(\beta) \right) \qquad (10.12)$$

which satisfy the following BC on the surface

$$\left(\frac{d}{d\beta} W(\beta) - qW(\beta) \right)_{\beta = \beta_s} = 0 \qquad (10.13)$$

and the radiation condition as x goes to infinity. Since the tangential component in (10.4) is negligible for a vertically polarized short dipole, only the normal component of the field can be considered in calculations.

10.2.4 Millington's Curve Fitting Approach

The Millington curve fitting method takes inhomogeneous path variations into account with the following recursive equations [52, 121] (see Fig. 10.2):

$$E_D = \sum_{k=1}^{N} E_k(s_k) - \sum_{k=2}^{N} E_k(s_{k-1}) \qquad (10.14)$$

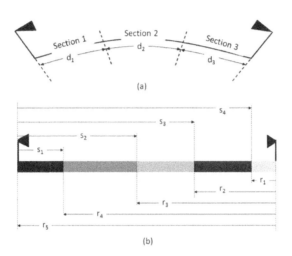

(a)

(b)

Figure 10.2 (a) A typical mixed path scenario on a 2D spherical Earth, (b) segments and related parameters.

$$s_k = \sum_{n=1}^{k} d_n = d_1 + d_2 + d_3 + ... + d_k \qquad (10.15)$$

$$E_R = \sum_{k=1}^{N} E_{N-k+1}(r_k) - \sum_{k=1}^{N-1} E_{N-k}(r_k) \qquad (10.16)$$

$$r_k = \sum_{n=1}^{k} d_{N-n+1} = d_N + d_{N-1} + d_{N-2} + ... + d_{N-k+1} \qquad (10.17)$$

where E_D and E_R are the direct fields using the paths s_k from the source to the receiver, and the reverse fields using the paths r_k from the receiver to the source, respectively (see Fig. 10.2). These fields are used to calculate the total field along the inhomogeneous propagation path as $E_T = (E_D + E_R)/2$.

The recursive equations in (10.14)–(10.17) reduce to

$$E_D = E_1(d_1) - E_2(d_1) + E_2(d_1 + d_2) \qquad (10.18)$$

$$E_R = E_2(d_2) - E_1(d_2) + E_1(d_2 + d_1) \qquad (10.19)$$

for a two-segment propagation path, where the field values $E_1(d_1)$, $E_1(d_2)$, $E_2(d_1)$, $E_2(d_2)$, $E_1(d_2 + d_1)$, and $E_2(d_1 + d_2)$ are defined as follows:

- $E_1(d_1)$ is the field strength at a distance d_1 over homogeneous Medium I

- $E_1(d_2)$ is the field strength at a distance d_2 over homogeneous Medium I

- $E_2(d_1)$ is the field strength at a distance d_1 over homogeneous Medium II

- $E_2(d_2)$ is the field strength at a distance d_2 over homogeneous Medium II

- $E_1(d_2 + d_1)$ is the field strength at a distance $d_1 + d_2$ over homogeneous Medium I

- $E_2(d_1 + d_2)$ is the field strength at a distance $d_1 + d_2$ over homogeneous Medium II

Similarly, they are given as

$$E_D = E_1(d_1) - E_2(d_1) + E_2(d_1 + d_2) - E_3(d_1 + d_2) + E_3(d_1 + d_2 + d_3) \quad (10.20)$$

$$E_R = E_3(d_3) - E_2(d_3 + d_2) - E_1(d_3 + d_2) + E_1(d_3 + d_2 + d_1) \qquad (10.21)$$

for a three-segment propagation path.

10.3 Numerical Surface-Wave Model

The FEMIX propagator uses the FEM-based standard PE. The accuracy of the numerical computations depends on the discretization. Several parameters, such as the frequency, source beamwidth, source tilt, source height, terrain profile, refractivity variations, need to be considered.

The FEMPE procedure of the FEMIX tool is applied as follows [44, 48]:

- An initial vertical field profile is injected into the FEMPE algorithm.

- An artificial lossy layer is used, and the vertical region is doubled to eliminate reflected waves coming from the top. In addition, a Tukey (tapered cosine) window is used for this purpose.

- The vertical field is propagated longitudinally by using the Crank–Nicolson approach.

- The new vertical field profile is taken as the initial field profile at the next range step.

- The procedure is repeated until the propagator reaches the desired range.

After calculating the field inside the entire computational domain, the PL can be obtained from [121]

$$PL = -20 \log |u| + 20 \log(4\pi) + 10 \log x - 30 \log \lambda. \qquad (10.22)$$

An inhomogeneous atmosphere is introduced through refractivity variations [3, 4]. Vertical refractivity variations ($n = n(x)$) may cause surface and/or elevated ducts. These are all implemented in the FEMIX model. Also, the Earth's curvature is included by replacing $n = n(x)$ with $n(x) + x/a_e$, where $a_e = 6,478$ km is the Earth's radius. It is customary to use the refractivity, $\left(N = (n(x) - 1) \times 10^6\right)$, or the modified refractivity, ($M = N + 157x$), with the height, x, given in kilometers. N is dimensionless but is measured in N units for convenience [4]. For the standard atmosphere, N decreases by about 40 N units/km, while M increases by about 117 N units/km (although the standard atmosphere defines an exponentially decreasing refractive index, this could be approximated as being linear for low heights) [44]. In other words, variations of the vertical gradient dM/dx of the modified refractivity are determined as atmospheric conditions as sub-refraction ($dM/dx > 117$ M units/km), standard ($dM/dx = 117$ M units/km), super-refraction ($dM/dx < 117$ M units/km), and ducting ($dM/dx < 0$).

A mathematical representation of all these cases can be achieved via

$$n(x) = n_0 \begin{cases} 1 + a_0 x & x \leq H_1 \\ 1 + a_0 H_1 + b_0 x & H_1 < x \leq H_2 \\ 1 + a_0 H_1 + b_0 H_2 + c_0 x & x > H_2 \end{cases} \qquad (10.23)$$

where n_0 is the refractivity at the surface. The user may form any kind of refractivity by specifying the first and second heights (in meters), H_1 and H_2, respectively, and the slopes a_0, b_0, and c_0 of the modified refractivity (supplied values are therefore multiplied by 10^{-9}, $a_0 \rightarrow a_0 \times 10^{-9}$). For example, a vertically linear decreasing refractivity up to 1 km height (i.e., a surface duct) can be specified in various ways: by giving (i) $H_1 = 0$ and $H_2 = 1000$, and $a_0 = 0$, $b_0 < 0$, c_0 is arbitrary; or (ii) $H_1 = 1000$ and $H_2 = 10000$, and $a_0 < 0$, b_0, and c_0 are arbitrary. Similarly, a bilinear refractivity up to 1 km (elevated duct) with a duct height at 600 m can be given by (i) $H_1 = 600$ and $H_2 = 1000$, and $a_0 > 0$, $b_0 < 0$, c_0 is arbitrary or, (ii) $H_1 = 0$ and $H_2 = 600$, and a_0 is arbitrary, $b_0 > 0$, $c_0 < 0$, etc.

10.4 FEMIX Package

The FEMIX package [125] is developed using MATLAB for the analysis and visualization of PL vs. range/height variations. The flow chart of the FEMIX model/package is given in Figs. 10.3 and 10.4. Note that the core FEMIX algorithms were all tested and validated before [2, 31, 34, 44, 48].

FEMIX has been designed in a way that the user may

- Visualize, load, and save terrain profiles over a maximum of six various paths.

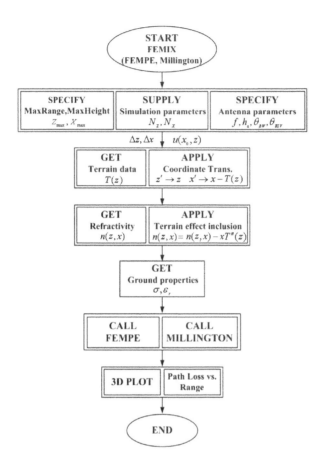

Figure 10.3   A flowchart of the FEMIX (Millington+FEMPE) GUI algorithm.

- Supply their input parameters.

- Be warned if inappropriate input values are entered.

- Plot an irregular terrain profile by locating a number of points and clicking the mouse, or the terrain profile may be supplied as an input file.

- Specify the terrain parameters such as permittivity, conductivity, and their lengths.

- Specify different refractivity profiles.

In order to simplify the use of the program, θ_{BW}, maximum height, and height step size are automatically specified with respect to the frequencies.

The main executable file of the package is FEMIX.EXE. Table 10.1 shows the input parameters of the package. The front panel is displayed in Fig. 10.5. The panel

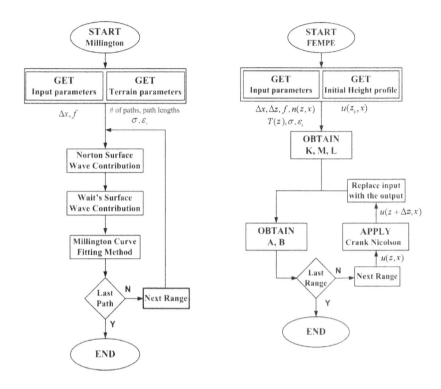

Figure 10.4   A flowchart of the Millington and FEMPE algorithms.

is mainly divided into two parts. The right part in Fig. 10.5 belongs to plot the PL calculations. The left part is for choosing the input parameters and the controls.

The left part is divided into three panels. The source frequency in mega hertz is supplied from the top panel. The range increment in kilometers for the horizontal grid width and range norm in kilometers for the calibration between numerical (FEM) and analytical (Millington) methods for a flat Earth are also entered here. The user may select one of the two methods or both. Plotting multiple PL results is also possible. The four operational pushbuttons (RUN, CLEAR, CLOSE, and INFO) are reserved for normal operations. Figures may be saved by ticking the Output TIFF box before running operation.

The middle panel is for the atmospheric refractivity. Three different linear refractivities may be selected by giving their heights (H_1 and H_2) and refractivity slopes (a_0, b_0, and c_0). The bottom panel is related to the terrain properties. The LOAD and SAVE buttons are used to load and record the irregular terrain data. By pressing the Locate Points button, new irregular terrain data are defined for each path on the bottom figure of the right part by using the number of points, range, and height values.

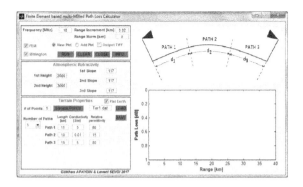

Figure 10.5   The front panel of the FEMIX GUI MATLAB package.

Table 10.1   User specified operational parameters of the FEMIX GUI package.

Operational Parameters	Explanation
Frequency [MHz]	Operating frequency
Range increment [m]	Range step on horizontal profile
Range norm [km]	Calibration range between Millington and FEMPE on horizontal profile
1st height [m]	Height of the first layer above terrain
2nd height [m]	Height of the second layer above terrain
1st slope [M/km]	Refractivity slope of the first layer
2nd slope [M/km]	Refractivity slope of the second layer
3rd slope [M/km]	Refractivity slope of the third layer
Number of points	Number of points to get terrain profile
Number of paths	Number of different paths
Length [km]	Path length of each path
Relative permittivity	Relative permittivity of each path
Conductivity [S/m]	Conductivity of each path

The electrical parameters of each path are also supplied by giving the conductivity and dielectric constant.

The top figure on the right side shows a mixed path scenario. PL vs. range results are displayed on the bottom plot. A 3D color field map showing fields vs. range–height is also plotted in a separate figure.

10.5   Characteristic Examples

Typical results obtained with the FEMIX package are presented in this part. The example given in Fig. 10.6 belongs to a three-path (sea–land–sea) 40 km flat Earth propagation scenario at three different frequencies (5 MHz, 10 MHz, and 20 MHz). A standard atmosphere is assumed. The electrical parameters of the sea (Path 1 and Path 3) and an island (Path 2) are $\sigma_g = 5$ S/m, $\varepsilon_r = 80$, and $\sigma_g = 0.01$ S/m, $\varepsilon_r = 15$, respectively. Both the scenario and parameters used in the simulations are shown in the figure. As seen, good agreement between the analytical model and FEM-based numerical model is obtained and PL increases while increasing frequency.

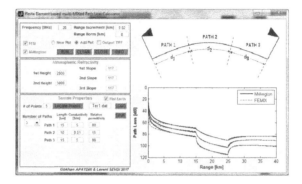

Figure 10.6 A three-path propagation scenario at 5 MHz, 10 MHz, and 20 MHz.

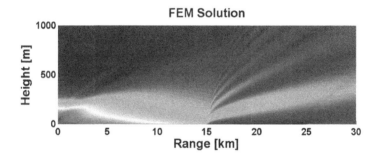

Figure 10.7 Fields vs. range/height as a 3D map at 20 MHz.

The fields vs. range/height are given as a 3D field map in Fig. 10.7, calculated via FEMIX at 20 MHz. Here, the effect of sea–land–sea transition is clearly observed.

A second example, given in Figs. 10.8–10.10, belongs to a three-path (sea–land–sea) propagation scenario for various frequencies. Again, a standard atmosphere is assumed. Here, FEMIX is compared with the Millington approach. The example belongs to PL vs. range variations over a three-segment 30 km mixed path. A 10 km long irregular terrain is 15 km away from the transmitter at three different frequencies. The electrical parameters of the sea (Path 1 and Path 3) and land path (Path 2) are $\sigma_g = 5$ S/m, $\varepsilon_r = 80$, and $\sigma_g = 0.01$ S/m, $\varepsilon_r = 15$, respectively. A user-plotted terrain profile is shown in Fig. 10.8.

Figure 10.9 shows PL vs. range for the scenario given in Fig. 10.8. The solid and dashed lines belong to the Millington and FEMPE models, respectively. As ob-

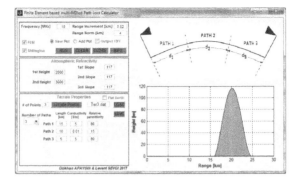

Figure 10.8 A three-path irregular propagation scenario.

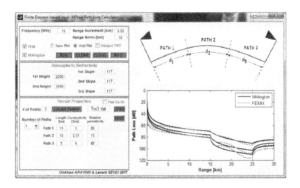

Figure 10.9 PL vs. range variations for 5 MHz, 10 MHz, and 15 MHz frequencies.

served, the PL increases with the frequency. Figure 10.10 shows fields vs. range/height obtained via FEMPE model for 5 MHz, 10 MHz, and 15 MHz, respectively. As seen, the energy tilt-up at the sea–land discontinuity and energy accumulation for different frequencies exist in front of the island. Note that in order to obtain a good resolution for the 3D map, the height increment is taken to be 10 m.

The third example, given in Figs. 10.11–10.13, belongs to a four-path (sea–land–sea–land) propagation scenario. The total length is 50 km. The electrical parameters of the sea (Path 1 and Path 3) and land paths (Path 2 and Path 4) are $\sigma_g = 5$ S/m, $\varepsilon_r = 80$, and $\sigma_g = 0.01$ S/m, $\varepsilon_r = 15$, respectively. Figure 10.11 shows the scenario and parameters used in the simulations. A user-plotted terrain profile is also shown.

Figure 10.12 shows PL vs. range for the scenario given in Fig. 10.11. The solid line belongs to the Millington model for the same four-path scenario (but without the irregular terrain) through a homogeneous atmosphere. The dashed line belongs to the FEMPE method, which takes both irregular terrain and atmospheric variations specified in Fig. 10.11 into account. Finally, Fig. 10.13 shows fields vs. range/height obtained via the FEMPE model.

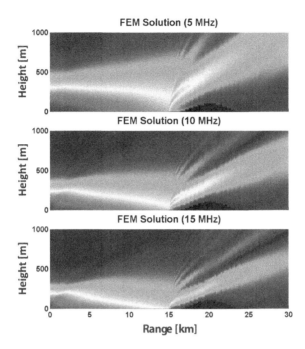

Figure 10.10   Three-dimensional fields vs. range/height map for 5 MHz, 10 MHz, and 15 MHz.

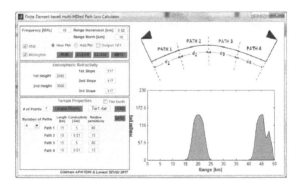

Figure 10.11   A four-path irregular propagation scenario.

Note that the physical and mathematical parameters are optimized in the FEMIX package for MF/LF/HF bands. The range step size is controlled by the user to consider the irregular terrain and the frequency. Although the large step size can affect the phase errors; the small steps increase the computation cost.

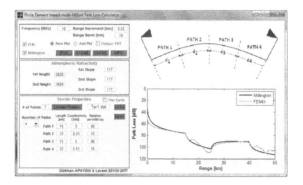

Figure 10.12   PL vs. range variations for the scenario in Fig. 10.11.

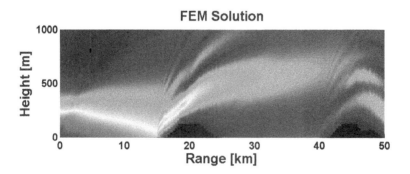

Figure 10.13   Three-dimensional fields vs. range/height map for the scenario in Fig. 10.12.

REFERENCES

1. L. Sevgi, "Guided waves and transverse fields: Transverse to what?," *IEEE Antennas and Propagation Magazine*, vol. 50, no. 6, pp. 221–225, Dec. 2008.

2. G. Apaydin and L. Sevgi, "Validation, verification and calibration in applied computational electromagnetics," *Applied Computational Electromagnetics Society Journal*, vol. 25, no. 12, pp. 1026–1035, Dec. 2010.

3. L. Sevgi, *Complex Electromagnetic Problems and Numerical Simulation Approaches*. John Wiley & Sons, 2003.

4. L. Sevgi, F. Akleman, and L. B. Felsen, "Groundwave propagation modeling: Problem-matched analytical formulations and direct numerical techniques," *IEEE Antennas and Propagation Magazine*, vol. 44, no. 1, pp. 55–75, Feb. 2002.

5. M. Levy, *Parabolic Equation Methods for Electromagnetic Wave Propagation*. IET, 2000.

6. L. Sevgi, "Numerical fourier transforms: Dft and fft," *IEEE Antennas and Propagation Magazine*, vol. 49, no. 3, pp. 238–243, Jun. 2007.

7. L. Sevgi, *Electromagnetic Modeling and Simulation*. John Wiley & Sons, 2014.

8. M. L. Meeks, "Radar propagation at low altitudes," *NASA STI/Recon Technical Report A*, vol. 83, p. 24897, 1982.

9. L. Sevgi, "Groundwave modeling and simulation strategies and path loss prediction virtual tools," *IEEE Transactions on Antennas and Propagation*, vol. 55, no. 6, pp. 1591–1598, Jun. 2007.

10. L. Sevgi, C. Uluisik, and F. Akleman, "A matlab-based two-dimensional parabolic equation radiowave propagation package," *IEEE Antennas and Propagation Magazine*, vol. 47, no. 4, pp. 164–175, Aug. 2005.

11. C. A. Tunc, F. Akleman, V. B. Erturk, A. Altintas, and L. Sevgi, "Fast integral equation solutions: Application to mixed path terrain profiles and comparisons with parabolic equation method," in *Complex Computing-Networks*, Springer, 2006, pp. 55-63.

12. F. Akleman and L. Sevgi, "A novel mom- and sspe-based groundwave-propagation field-strength prediction simulator," *IEEE Antennas and Propagation Magazine*, vol. 49, no. 5, pp. 69–82, Oct. 2007.

13. F. Akleman and L. Sevgi, "A novel finite-difference time-domain wave propagator," *IEEE Transactions on Antennas and Propagation*, vol. 48, no. 5, pp. 839–841, May 2000.

14. M. O. Ozyalcin, F. Akleman, and L. Sevgi, "A novel tlm-based time-domain wave propagator," *IEEE Transactions on Antennas and Propagation*, vol. 51, no. 7, pp. 1679–1680, Jul. 2003.

15. L. Sevgi, "A mixed-path groundwave field-strength prediction virtual tool for digital radio broadcast systems in medium and short wave bands," *IEEE Antennas and Propagation Magazine*, vol. 48, no. 4, pp. 19–27, Aug. 2006.

16. L. Sevgi and L. B. Felsen, "A new algorithm for ground wave propagation based on a hybrid ray-mode approach," *International Journal of Numerical Modelling: Electronic Networks, Devices and Fields*, vol. 11, no. 2, pp. 87–103, Mar. 1998.

17. L. Sevgi, "A ray-shooting visualization matlab package for 2d ground-wave propagation simulations," *IEEE Antennas and Propagation Magazine*, vol. 46, no. 4, pp. 140–145, Aug. 2004.

18. M. A. Leontovich and V. A. Fock, "Solution of the problem of propagation of electromagnetic waves along the earth's surface by the method of parabolic equation," *Academy of Sciences of the USSR: Journal of Physics*, vol. 10, pp. 13–23, 1946.

19. V. A. Fock, *Electromagnetic Diffraction and Propagation Problems*. Pergamon, 1965.

20. F. D. Tappert, "The parabolic approximation method," in *Wave Propagation and Underwater Acoustics*, Springer, 1977, pp. 224-287.

21. I. Sirkova and M. Mikhalev, "Parabolic wave equation method applied to the tropospheric ducting propagation problem: A survey," *Electromagnetics*, vol. 26, no. 2, pp. 155–173, 2006.

22. D. H. Huang, "Finite-element solution to the parabolic wave-equation," *Journal of the Acoustical Society of America*, vol. 84, no. 4, pp. 1405–1413, Oct. 1988.

23. V. M. Deshpande and M. D. Deshpande, "Study of electromagnetic wave propagation through dielectric slab doped randomly with thin metallic wires using finite element method," *IEEE Microwave and Wireless Components Letters*, vol. 15, no. 5, pp. 306–308, May 2005.

24. K. Arshad and F. Katsriku, "An investigation of tropospheric wave propagation using finite elements," *WSEAS Transactions on Communications*, vol. 4, no. 11, pp. 1186–1192, 2005.

25. M. D. Feit and J. A. Fleck, "Light propagation in graded-index optical fibers," *Applied Optics*, vol. 17, no. 24, pp. 3990–3998, 1978.

26. D. J. Thomson and N. R. Chapman, "A wide-angle split-step algorithm for the parabolic equation," *Journal of the Acoustical Society of America*, vol. 74, no. 6, pp. 1848–1854, 1983.

27. G. Apaydin, O. Ozgun, M. Kuzuoglu, and L. Sevgi, "A novel two-way finite-element parabolic equation groundwave propagation tool: Tests with canonical structures and calibration," *IEEE Transactions on Geoscience and Remote Sensing*, vol. 49, no. 8, pp. 2887–2899, Aug. 2011.

28. B. M. Hannah, *Modelling and Simulation of GPS Multipath Propagation*. PhD thesis, Queensland University of Technology, 2001.

29. J. R. Wait, "The scope of impedance boundary-conditions in radio propagation," *IEEE Transactions on Geoscience and Remote Sensing*, vol. 28, no. 4, pp. 721–723, Jul. 1990.

30. M. F. Levy, "Horizontal parabolic equation solution of radiowave propagation problems on large domains," *IEEE Transactions on Antennas and Propagation*, vol. 43, no. 2, pp. 137–144, Feb. 1995.

31. G. Apaydin and L. Sevgi, "The split-step-fourier and finite-element-based parabolic-equation propagation-prediction tools: Canonical tests, systematic comparisons, and calibration," *IEEE Antennas and Propagation Magazine*, vol. 52, no. 3, pp. 66–79, Jun. 2010.

32. G. D. Dockery and J. R. Kuttler, "An improved impedance-boundary algorithm for fourier split-step solutions of the parabolic wave equation," *IEEE Transactions on Antennas and Propagation*, vol. 44, no. 12, pp. 1592–1599, Dec. 1996.

33. J. R. Kuttler and R. Janaswamy, "Improved fourier transform methods for solving the parabolic wave equation," *Radio Science*, vol. 37, no. 2, pp. 1–11, 2002.

34. G. Apaydin and L. Sevgi, "A novel split-step parabolic-equation package for surface-wave propagation prediction along multiple mixed irregular-terrain paths," *IEEE Antennas and Propagation Magazine*, vol. 52, no. 4, pp. 90–97, Aug. 2010.

35. J.-M. Jin, *The Finite Element Method in Electromagnetics*. John Wiley & Sons, 2015.

36. J. L. Volakis, A. Chatterjee, and L. C. Kempel, *Finite Element Method Electromagnetics: Antennas, Microwave Circuits, and Scattering Applications*, vol. 6. John Wiley & Sons, 1998.

37. W. Y. Yang, W. Cao, T.-S. Chung, and J. Morris, *Applied Numerical Methods Using MATLAB*. John Wiley & Sons, 2005.

38. L. Barclay, *Propagation of Radiowaves*. IET, 2003.

39. D. E. Kerr, *Propagation of Short Radio Waves*. IET, 1951.

40. R. J. Luebbers, "Propagation prediction for hilly terrain using gtd wedge diffraction," *IEEE Transactions on Antennas and Propagation*, vol. 32, no. 9, pp. 951–955, Sep. 1984.

41. J. H. Whitteker, "Fresnel-kirchhoff theory applied to terrain diffraction problems," *Radio Science*, vol. 25, no. 5, pp. 837–851, 1990.

42. C. A. Tunc, A. Altintas, and V. B. Erturk, "Examination of existent propagation models over large inhomogeneous terrain profiles using fast integral equation solution," *IEEE Transactions on Antennas and Propagation*, vol. 53, no. 9, pp. 3080–3083, Sep. 2005.

43. H. Oraizi and S. Hosseinzadeh, "Radio-wave-propagation modeling in the presence of multiple knife edges by the bidirectional parabolic-equation method," *IEEE Transactions on Vehicular Technology*, vol. 56, no. 3, pp. 1033–1040, May 2007.

44. G. Apaydin and L. Sevgi, "Fem-based surface wave multimixed-path propagator and path loss predictions," *IEEE Antennas and Wireless Propagation Letters*, vol. 8, pp. 1010–1013, 2009.

45. G. Apaydin and L. Sevgi, "A canonical test problem for computational electromagnetics (cem): Propagation in a parallel-plate waveguide," *IEEE Antennas and Propagation Magazine*, vol. 54, no. 4, pp. 290–315, Aug. 2012.

46. G. Apaydin and L. Sevgi, "Method of moment (mom) modeling for resonating structures: Propagation inside a parallel plate waveguide," *Applied Computational Electromagnetics Society Journal*, vol. 27, no. 10, pp. 842–849, Oct. 2012.

47. G. Apaydin and L. Sevgi, "Groundwave propagation at short ranges and accurate source modeling," *IEEE Antennas and Propagation Magazine*, vol. 55, no. 3, pp. 244–262, Jun. 2013.

48. G. Apaydin and L. Sevgi, "Numerical investigations of and path loss predictions for surface wave propagation over sea paths including hilly island transitions," *IEEE Transactions on Antennas and Propagation*, vol. 58, no. 4, pp. 1302–1314, Apr. 2010.

49. G. Apaydin and L. Sevgi, "Propagation modeling and path loss prediction tools for high frequency surface wave radars," *Turkish Journal of Electrical Engineering and Computer Sciences*, vol. 18, no. 3, pp. 469–484, 2010.

50. K. A. Norton, "The propagation of radio waves over the surface of the earth and in the upper atmosphere," *Proceedings of the Institute of Radio Engineers*, vol. 24, no. 10, pp. 1367–1387, Oct. 1936.

51. J. R. Wait, *Electromagnetic Waves in Stratified Media: Revised Edition Including Supplemented Material*. Elsevier, 2013.

52. G. Millington, "Ground-wave propagation over an inhomogeneous smooth earth," *Proceedings of the IEE-Part III: Radio and Communication Engineering*, vol. 96, no. 39, pp. 53–64, Jan. 1949.

53. L. Sevgi, "A numerical millington propagation package for medium and short wave drm systems field strength predictions," *IEEE Broadcast Technology Society Newsletter*, vol. 14, no. 3, pp. 9–11, 2006.

54. K. Furutsu, "A systematic theory of wave propagation over irregular terrain," *Radio Science*, vol. 17, no. 5, pp. 1037–1050, 1982.

55. R. Ott, "A new method for predicting hf ground wave attenuation over inhomogeneous, irregular terrain," *DTIC Document*, 1971.

56. S. W. Marcus, "A hybrid (finite difference-surface green's function) method for computing transmission losses in an inhomogeneous atmosphere over irregular terrain," *IEEE Transactions on Antennas and Propagation*, vol. 40, no. 12, pp. 1451–1458, Dec. 1992.

57. R. Janaswamy, "A fredholm integral equation method for propagation predictions over small terrain irregularities," *IEEE Transactions on Antennas and Propagation*, vol. 40, no. 11, pp. 1416–1422, Nov. 1992.

58. R. Janaswamy, "A fast finite difference method for propagation predictions over irregular, inhomogeneous terrain," *IEEE Transactions on Antennas and Propagation*, vol. 42, no. 9, pp. 1257–1267, Sep. 1994.

59. J. T. Hviid, J. B. Andersen, J. Toftgard, and J. Bojer, "Terrain-based propagation model for rural area-an integral equation approach," *IEEE Transactions on Antennas and Propagation*, vol. 43, no. 1, pp. 41–46, Jan. 1995.

60. K. H. Craig, "Propagation modelling in the troposphere: Parabolic equation method," *Electronics Letters*, vol. 24, no. 18, pp. 1136–1139, 1988.

61. A. E. Barrios, "Parabolic equation modeling in horizontally inhomogeneous environments," *IEEE Transactions on Antennas and Propagation*, vol. 40, no. 7, pp. 791–797, Jul. 1992.

62. J. R. Kuttler and G. D. Dockery, "Theoretical description of the parabolic approximation/fourier split-step method of representing electromagnetic propagation in the troposphere," *Radio Science*, vol. 26, no. 2, pp. 381–393, 1991.

63. D. J. Donohue and J. R. Kuttler, "Propagation modeling over terrain using the parabolic wave equation," *IEEE Transactions on Antennas and Propagation*, vol. 48, no. 2, pp. 260–277, Feb. 2000.

64. M. D. Collins and R. B. Evans, "A two-way parabolic equation for acoustic backscattering in the ocean," *Journal of the Acoustical Society of America*, vol. 91, no. 3, pp. 1357–1368, 1992.

65. M. D. Collins, "A two-way parabolic equation for elastic media," *Journal of the Acoustical Society of America*, vol. 93, no. 4, pp. 1815–1825, 1993.

66. J. F. Lingevitch, M. D. Collins, M. J. Mills, and R. B. Evans, "A two-way parabolic equation that accounts for multiple scattering," *Journal of the Acoustical Society of America*, vol. 112, no. 2, pp. 476–480, 2002.

67. M. J. Mills, M. D. Collins, and J. F. Lingevitch, "Two-way parabolic equation techniques for diffraction and scattering problems," *Wave Motion*, vol. 31, no. 2, pp. 173–180, 2000.

68. O. Ozgun, "Recursive two-way parabolic equation approach for modeling terrain effects in tropospheric propagation," *IEEE Transactions on Antennas and Propagation*, vol. 57, no. 9, pp. 2706–2714, Sep. 2009.

69. C. A. Balanis, *Advanced Engineering Electromagnetics*. John Wiley & Sons, 2012.

70. L. B. Felsen and N. Marcuvitz, *Radiation and Scattering of Waves*, vol. 31. John Wiley & Sons, 1994.

71. D. G. Dudley, *Mathematical Foundations for Electromagnetic Theory*. IEEE Press, 1994.

72. M. N. O. Sadiku, *Numerical Techniques in Electromagnetics with MATLAB*. CRC Press, 2011.

73. T. Rylander, P. Ingelstrom, and A. Bondeson, *Computational Electromagnetics*. New York: Springer, 2013.

74. A. F. Peterson, S. L. Ray, and R. Mittra, *Computational Methods for Electromagnetics*, vol. 2. New York: IEEE Press, 1998.

75. J.-M. Jin, *Theory and Computation of Electromagnetic Fields*. John Wiley & Sons, 2011.

76. R. Garg, *Analytical and Computational Methods in Electromagnetics*. Artech House, 2008.

77. R. C. Hansen, *Geometric Theory of Diffraction*. IEEE Press, 1981.

78. R. G. Kouyoumjian and P. H. Pathak, "A uniform geometrical theory of diffraction for an edge in a perfectly conducting surface," *Proceedings of the IEEE*, vol. 62, no. 11, pp. 1448–1461, Nov. 1974.

79. P. Y. Ufimtsev, *Fundamentals of the Physical Theory of Diffraction*. Hoboken, NJ: John Wiley & Sons, Inc., 2014.

80. P. Y. Ufimtsev, *Theory of Edge Diffraction in Electromagnetics. Origination and Validation of the Physical Theory of Diffraction*. Raleigh, NC: Revised printing, SciTech Publishing Inc., 2009.

81. F. Hacivelioglu, L. Sevgi, and P. Y. Ufimtsev, "Electromagnetic wave scattering from a wedge with perfectly reflecting boundaries: Analysis of asymptotic techniques," *IEEE Antennas and Propagation Magazine*, vol. 53, no. 3, pp. 232–253, Jun. 2011.

82. F. Hacivelioglu, M. A. Uslu, and L. Sevgi, "A matlab-based virtual tool for the electromagnetic wave scattering from a perfectly reflecting wedge," *IEEE Antennas and Propagation Magazine*, vol. 53, no. 6, pp. 234–243, Dec. 2011.

83. L. Sevgi, "Sturm-liouville equation: The bridge between eigenvalue and green's function problems," *Turkish Journal of Electrical Engineering and Computer Sciences*, vol. 14, no. 2, pp. 293–311, 2006.

84. R. F. Harrington and J. L. Harrington, *Field Computation by Moment Methods*. Oxford University Press, 1996.

85. K. S. Kunz and R. J. Luebbers, *The Finite Difference Time Domain Method for Electromagnetics*. CRC Press, 1993.

86. D. M. Sullivan, *Electromagnetic Simulation Using the FDTD Method*. John Wiley & Sons, 2013.

87. A. Taflove and S. C. Hagness, *Computational Electrodynamics*. Artech House, 2005.

88. C. Christopoulos, *The Transmission-Line Modeling Method*. Oxford Press, 1995.

89. P. Russer, *Electromagnetics, Microwave Circuit and Antenna Design for Communications Engineering*. Artech House, 2003.

90. L. Felsen and A. Kamel, "Hybrid ray-mode formulation of parallel plane waveguide green's functions," *IEEE Transactions on Antennas and Propagation*, vol. 29, no. 4, pp. 637–649, Jul. 1981.

91. L. B. Felsen, F. Akleman, and L. Sevgi, "Wave propagation inside a two-dimensional perfectly conducting parallel-plate waveguide: Hybrid ray-mode techniques and their visualizations," *IEEE Antennas and Propagation Magazine*, vol. 46, no. 6, pp. 69–89, Dec. 2004.

92. L. Sevgi, F. Akleman, and L. B. Felsen, "Visualizations of wave dynamics in a wedge waveguide with non-penetrable boundaries: Normal-, adiabatic-, and intrinsic-mode representations," *IEEE Antennas and Propagation Magazine*, vol. 49, no. 3, pp. 76–94, Jun. 2007.

93. G. Apaydin and L. Sevgi, "Two-way propagation modeling in waveguides with three-dimensional finite-element and split-step fourier-based pe approaches," *IEEE Antennas and Wireless Propagation Letters*, vol. 10, pp. 975–978, 2011.

94. R. Ding, L. Tsang, and H. Braunisch, "Wave propagation in a randomly rough parallel-plate waveguide," *IEEE Transactions on Microwave Theory and Techniques*, vol. 57, no. 5, pp. 1216–1223, May 2009.

95. E. Arvas and L. Sevgi, "A tutorial on the method of moments," *IEEE Antennas and Propagation Magazine*, vol. 54, no. 3, pp. 260–275, Jun. 2012.

96. O. Ozgun, G. Apaydin, M. Kuzuoglu, and L. Sevgi, "Petool: Matlab-based one-way and two-way split-step parabolic equation tool for radiowave propagation over variable terrain," *Computer Physics Communications*, vol. 182, no. 12, pp. 2638–2654, 2011.

97. G. Apaydin and L. Sevgi, "Calibration of three-dimensional parabolic-equation propagation models with the rectangular waveguide problem," *IEEE Antennas and Propagation Magazine*, vol. 54, no. 6, pp. 102–116, Dec. 2012.

98. R. Martelly and R. Janaswamy, "An adi-pe approach for modeling radio transmission loss in tunnels," *IEEE Transactions on Antennas and Propagation*, vol. 57, no. 6, pp. 1759–1770, Jun. 2009.

99. J. B. Keller, "Geometrical theory of diffraction," *Journal of the Optical Society of America*, vol. 52, pp. 116–130, 1962.

100. C. A. Zelley and C. C. Constantinou, "A three-dimensional parabolic equation applied to vhf/uhf propagation over irregular terrain," *IEEE Transactions on Antennas and Propagation*, vol. 47, no. 10, pp. 1586–1596, Oct. 1999.

101. A. V. Popov and N. Y. Zhu, "Modeling radio wave propagation in tunnels with a vectorial parabolic equation," *IEEE Transactions on Antennas and Propagation*, vol. 48, no. 9, pp. 1403–1412, Sep. 2000.

102. R. Martelly and R. Janaswamy, "Modeling radio transmission loss in curved, branched and rough-walled tunnels with the adi-pe method," *IEEE Transactions on Antennas and Propagation*, vol. 58, no. 6, pp. 2037–2045, Jun. 2010.

103. A. Yagbasan, C. A. Tunc, V. B. Erturk, A. Altintas, and R. Mittra, "Characteristic basis function method for solving electromagnetic scattering problems over rough terrain profiles," *IEEE Transactions on Antennas and Propagation*, vol. 58, no. 5, pp. 1579–1589, May 2010.

104. F. Akleman and L. Sevgi, "Time and frequency domain wave propagators," *Applied Computational Electromagnetics Society Journal*, vol. 15, no. 3, pp. 186–208, 2000.

105. F. Akleman and L. Sevgi, "Realistic surface modeling for a finite-difference time-domain wave propagator," *IEEE Transactions on Antennas and Propagation*, vol. 51, no. 7, pp. 1675–1679, Jul. 2003.

106. K. G. Budden, *The Waveguide Mode Theory of Wave Propagation*. London: Logos Press, 1961.

107. R. H. Hardin and F. D. Tappert, "Applications of the split-step fourier method to the numerical solution of nonlinear and variable coefficient wave equations," *Society for Industrial and Applied Mathematics Review*, vol. 15, no. 2, p. 423, 1973.

108. G. D. Dockery, "Modeling electromagnetic wave propagation in the troposphere using the parabolic equation," *IEEE Transactions on Antennas and Propagation*, vol. 36, no. 10, pp. 1464–1470, Oct. 1988.

109. P. D. Holm, "Wide-angle shift-map pe for a piecewise linear terrain-a finite-difference approach," *IEEE Transactions on Antennas and Propagation*, vol. 55, no. 10, pp. 2773–2789, Oct. 2007.

110. C. Mias, "Fast computation of the nonlocal boundary condition in finite difference parabolic equation radiowave propagation simulations," *IEEE Transactions on Antennas and Propagation*, vol. 56, no. 6, pp. 1699–1705, Jun. 2008.

111. M. F. Levy, "Parabolic equation modelling of propagation over irregular terrain," *Electronics Letters*, vol. 26, no. 15, pp. 1153–1155, Jul. 1990.

112. A. E. Barrios, "A terrain parabolic equation model for propagation in the troposphere," *IEEE Transactions on Antennas and Propagation*, vol. 42, no. 1, pp. 90–98, Jan. 1994.

113. R. Janaswamy, "A curvilinear coordinate-based split-step parabolic equation method for propagation predictions over terrain," *IEEE Transactions on Antennas and Propagation*, vol. 46, no. 7, pp. 1089–1097, Jul. 1998.

114. T.-K. Tsay, B. A. Ebersole, and P. L.-F. Liu, "Numerical modelling of wave propagation using parabolic approximation with a boundary-fitted co-ordinate system," *International Journal for Numerical Methods in Engineering*, vol. 27, no. 1, pp. 37–55, 1989.

115. M. F. Levy and A. A. Zaporozhets, "Parabolic equation techniques for scattering," *Wave Motion*, vol. 31, no. 2, pp. 147–156, 2000.

116. O. Ozgun, G. Apaydin, M. Kuzuoglu, and L. Sevgi, "Two-way fourier split step algorithm over variable terrain with narrow and wide angle propagators," in *2010 IEEE Antennas and Propagation Society International Symposium*, pp. 1–4, Jul. 2010.

117. M. F. Levy, "Diffraction studies in urban environment with wide-angle parabolic equation method," *Electronics Letters*, vol. 28, no. 16, pp. 1491–1492, Jul. 1992.

118. J. R. Kuttler, "Differences between the narrow-angle and wide-angle propagators in the split-step fourier solution of the parabolic wave equation," *IEEE Transactions on Antennas and Propagation*, vol. 47, no. 7, pp. 1131–1140, Jul. 1999.

119. J. F. Claerbout, *Fundamentals of Geophysical Data Processing with Application to Petroleum Prospect*. New York: McGraw-Hill, 1976.

120. ITU-R, "Planning parameters for digital sound broadcasting at frequencies below 30 mhz," *International Telecommunications Union-Radiocommunication Recommendation BS.1615*, Jun. 2003.

121. ITU-R, "Groundwave propagation curves for frequencies between 10 khz and 30 mhz," *International Telecommunications Union-Radiocommunication Recommendation P.368-9*, Feb. 2007.

122. L. Sevgi, A. Ponsford, and H. C. Chan, "An integrated maritime surveillance system based on high-frequency surface-wave radars. 1. Theoretical background and numerical simulations," *IEEE Antennas and Propagation Magazine*, vol. 43, no. 4, pp. 28–43, Aug. 2001.

123. A. M. Ponsford, L. Sevgi, and H. C. Chan, "An integrated maritime surveillance system based on high-frequency surface-wave radars. 2. Operational status and system performance," *IEEE Antennas and Propagation Magazine*, vol. 43, no. 5, pp. 52–63, Oct. 2001.

124. L. Sevgi, "Novel digital broadcast/communication systems and groundwave propagation prediction requirements," in *2006 First European Conference on Antennas and Propagation*, pp. 1–5, Nov. 2006.

125. G. Apaydin and L. Sevgi, "Matlab-based fem-parabolic-equation tool for path-loss calculations along multi-mixed-terrain paths," *IEEE Antennas and Propagation Magazine*, vol. 56, no. 3, pp. 221–236, Jun. 2014.

Index

Radio Wave Propagation and Parabolic Equation Modeling, First Edition. By Gökhan Apaydin, Levent Sevgi
© 2017 by the Institute of Electrical and Electronic Engineers, Inc. Published 2017 by John Wiley & Sons, Inc.